AUTODESK® DRAINAGE DESIGN FOR INFRAWORKS 360™

ESSENTIALS

AUTODESK® DRAINAGE DESIGN FOR INFRAWORKS 360™

ESSENTIALS

Eric Chappell

AUTODESK. Official Press

SYBEX® A Wiley Brand

Senior Acquisitions Editor: Willem Knibbe
Development Editor: Sara Barry
Technical Editor: Tom Richardson
Production Editor: Eric Charbonneau
Copy Editor: Kim Wimpsett
Editorial Manager: Pete Gaughan
Vice President and Executive Group Publisher: Richard Swadley
Associate Publisher: Chris Webb
Book Designer: Happenstance Type-O-Rama
Compositor: Cody Gates, Happenstance Type-O-Rama
Proofreader: Rebecca Rider
Indexer: Ted Laux
Project Coordinator, Cover: Todd Klemme
Cover Designer: Wiley
Cover Image: Eric Chappell

ISBN: 978-1-118-91596-7 (pbk.); 978-1-118-90178-6 (ebk.); 978-1-118-91598-1 (ebk.)

10 9 8 7 6 5 4 3 2 1

Dear Reader,

Thank you for choosing *Autodesk Drainage Design for Infraworks 360 Essentials*. This book is part of a family of premium-quality Sybex books, all of which are written by outstanding authors who combine practical experience with a gift for teaching.

Sybex was founded in 1976. More than 30 years later, we're still committed to producing consistently exceptional books. With each of our titles, we're working hard to set a new standard for the industry. From the paper we print on to the authors we work with, our goal is to bring you the best books available.

I hope you see all that reflected in these pages. I'd be very interested to hear your comments and get your feedback on how we're doing. Feel free to let me know what you think about this or any other Sybex book by sending me an email at `contactus@sybex.com`. If you think you've found a technical error in this book, please visit `http://sybex.custhelp.com`. Customer feedback is critical to our efforts at Sybex.

Best regards,

CHRIS WEBB
Associate Publisher
Sybex, an Imprint of Wiley

To the Sonz

Acknowledgments

Another mini-book! What a blast it has been to write about the advanced modules of InfraWorks, and this book about Autodesk® Drainage Design for InfraWorks 360™ has been no exception. It really is mind-blowing what this technology can do and how good it looks while it's doing it.

As with the other InfraWorks books that have come out this year, I have to thank Willem Knibbe for making this book even happen in the first place. And again I would like to thank Pete Gaughan for stepping in, taking the reins, and being helpful, decisive, positive, and just wonderful to work with. Tom Richardson was technical editor for this and all the other InfraWorks books, and I can't think of anyone else I'd rather have when it comes to providing a thorough and insightful technical working over. And last but certainly not least, thanks to Sara Barry and Eric Charbonneau for your tireless efforts correcting my writing and making it so much better.

I say it in every book and it's true every time, my family takes the brunt of the condition that is book writing. I hope that all of the time and effort is worth it and that this book, like the others, will serve as an example of quality work to my children and a source of pride for my wife.

ABOUT THE AUTHOR

Eric Chappell has been working, teaching, writing, and consulting in the world of civil engineering software for more than 20 years, and he is a recognized expert in the world of Autodesk® InfraWorks 360™ software. In September 2013, Eric joined the Autodesk family as a Premium Services Specialist. In the 12 years prior to that, he wrote training materials and performed training for end users, trainers, and Autodesk employees around the globe. For several years, he has worked with Autodesk in authoring and developing two Autodesk certification exams. He also served as design systems manager for Timmons Group, a civil engineering and surveying firm based in Richmond, Virginia, where he managed software, standards, and training for more than 200 users. Eric is also a highly rated instructor at Autodesk University, where he has taught for the past 10 years.

Prior to writing and consulting, Eric spent nearly 10 years in the civil engineering and surveying fields while working for the H.F. Lenz Company in Johnstown, Pennsylvania. During his time at H.F. Lenz, he gained considerable practical experience as a survey crewman, designer, engineer, and CAD supervisor. Eric also holds a B.S. degree in civil engineering technology from the University of Pittsburgh at Johnstown and is certified in Pennsylvania as an EIT.

Eric is originally from southwestern Pennsylvania, but he has lived in the Richmond, Virginia, area for the past 13 years with his wife and four children. He enjoys being outdoors and spending time with his family. He can sometimes be seen playing drums for the band Sons of Zebedee, which plays at a variety of events in central Virginia.

If you would like to contact the author regarding comments or suggestions, please email `InfraWorksEssentials@gmail.com`. You are also welcome to visit Eric's blog at `http://ericchappell.blogspot.com`.

Contents at a Glance

Contents

INTRODUCTION

I can't imagine writing a book or learning a new program in a more exciting time. We in the civil engineering, planning, construction, geospatial, and other related fields are experiencing the "changing of the guard," a new era in design software technology. The Autodesk® AutoCAD® Civil 3D® software is still the powerful flagship that is a global monument to civil engineering design, but there is a new kid on the block—and he's quick, smart, and attractive, and he's a great communicator. His name is Autodesk® InfraWorks™.

InfraWorks has already begun revolutionizing the way design is done, and one of its greatest attributes is that it is simply fun to use—right up there with playing a video game. And who wouldn't want to be paid to play a video game all day? But in all seriousness, not only is InfraWorks 360 fun, it gets the job done.

When I realized that I was going to be one of the first people to write an InfraWorks book, I was of course overjoyed! But then I also felt a great weight of responsibility. If I was going to be partially responsible for educating the world about this exciting new industry-changing product, I needed to do it right. My panic attack was brief, however, because I realized that I had done this before, having written four editions of *Autodesk Civil 3D Essentials*. I am quite pleased and proud of the quality, reception, feedback, and success of those books, and I intend to use a similar approach for this one. That approach was built upon several basic criteria that I am now applying to *Autodesk Drainage Design for InfraWorks 360 Essentials*.

- It should be basic enough to enable anyone to learn Drainage Design for InfraWorks 360.
- It should be in-depth enough to enable a person to be productive using Drainage Design for InfraWorks 360 for basic tasks.
- It should foster understanding by associating the things you do in Drainage Design for InfraWorks 360 with familiar things you see every day.
- The examples and exercises should be based on the real world.
- The book should not simply demonstrate random software features but should teach the process of project completion using Drainage Design for InfraWorks 360.

While writing each chapter of this book, I have tried to meet these criteria. I hope before opening this book, you have gone through *Autodesk InfraWorks and InfraWorks 360 Essentials* and have become familiar with the concept of InfraWorks, its user interface, and its environment. In addition, I hope you have completed the *Autodesk Roadway Design for InfraWorks 360 Essentials* and *Autodesk Bridge Design for InfraWorks 360 Essentials* ebooks. Just as Drainage Design for InfraWorks 360 is built upon InfraWorks 360, this book is built upon its counterparts. In addition, if you've experienced any of the aforementioned InfraWorks books, then you're also familiar with the Bimsville Bypass project. In this book, you'll create, modify, and analyze watersheds, culverts, and drainage networks along the Bimsville Bypass and neighboring proposed industrial park.

The Full InfraWorks Essentials Library

This ebook is part of a complete library covering InfraWorks. All are available at Sybex.com and through a variety of online resellers.

Autodesk InfraWorks and InfraWorks 360 Essentials
Available in print and as an ebook

Autodesk Roadway Design for InfraWorks 360 Essentials
Available as an ebook

Autodesk Bridge Design for InfraWorks 360 Essentials
Available as an ebook

Autodesk Drainage Design for InfraWorks 360 Essentials
Available as an ebook

Who Should Read This Book

This book should be read by anyone who already understands the core InfraWorks features as well as the Autodesk® Roadway Design for InfraWorks 360™ features and is ready to move on to drainage design. It is designed to be used in a formal classroom environment, in your home office learning on your own, and everywhere in between. It is appropriate for ages ranging from high school to retirement, and although it is intended for those who have no experience or skill with Drainage Design for InfraWorks 360, it can also serve as a great resource for refreshing one's knowledge base or for filling in any gaps.

Here are some specific examples of individuals who would benefit from reading this book:

- Anyone involved in the planning or preliminary design stages of infrastructure or land development projects
- High-school students following a design-related educational track
- College students learning to be designers, engineers, or GIS professionals
- Employees who have recently joined a company that utilizes InfraWorks 360
- Employees who work for companies that have recently implemented InfraWorks 360
- Experienced InfraWorks 360 users who are self-taught and who want to fill in gaps in their knowledge base
- Civil 3D users

What You Will Learn

The goal of this book is to make you productive with Drainage Design for InfraWorks 360 in a relatively short time. It begins by giving you a detailed tour of the user interface and the environment you will be working in. It then covers nearly all of the Drainage Design for InfraWorks 360 features using relatively simple but realistic examples. It takes you through the progression of a project (the Bimsville Bypass) so that you can learn the logic of how Drainage Design for InfraWorks 360 is applied to make projects more successful.

After completing this book, you will be able to use Drainage Design for InfraWorks 360 to create, modify, and analyze preliminary drainage designs in your InfraWorks models. You will be able to work on a team that also uses Drainage Design for InfraWorks 360, and you will be able to create and share drainage design models with them.

Free Autodesk Software for Students and Educators

The Autodesk Education Community is an online resource with more than 5 million members that enables educators and students to download— for free (see website for terms and conditions)—the same software used by professionals worldwide. You can also access additional tools and materials to help you design, visualize, and simulate ideas. Connect with other learners to stay current with the latest industry trends and get the most out of your designs. Get started today at `www.autodesk.com/joinedu`.

What You Need

The following sections highlight the key things you will need to successfully complete the entire book.

Software

To work through the exercises in this book, you will need to have InfraWorks 360 installed on your computer, with Drainage Design for InfraWorks 360 enabled. You should refer to the Autodesk website (`www.autodesk.com`) and make sure your computer's specifications are equal or better than the recommended system requirements for the current release of these products.

An InfraWorks 360 Account

For Drainage Design for InfraWorks 360 to be enabled, you must have InfraWorks 360 installed. Autodesk InfraWorks (not 360) does not have the ability to accept Drainage Design for InfraWorks as an add-on program. You must also be entitled to Drainage Design for InfraWorks 360 either as a trial version or through your Autodesk account. If you work for a company or organization, check with a software manager within your organization about getting set up with the proper software and entitlements. Or, if you do not have such a person in your organization, contact your Autodesk reseller. You can also visit `www.autodesk.com/infraworks` for more information.

Access to Cloud Credits

There are several tasks in the exercises that will require cloud credits for you to complete. Check with your software administrator or with an Autodesk representative to ensure that you have access to cloud credits. If you do not, you can skip those portions of the exercises and pick back up at the following exercise. The need for cloud credits impacts only a small portion of the book.

Exercise Files

To complete the exercises, you will need to download the necessary files from `www.sybex.com/go/drainageessentials`. Here you will find a list of ZIP files, one for each chapter, which you should unzip to the local C: drive of your computer. This will create a folder named `InfraWorks Drainage Essentials` with the chapter folder inside it. As you unzip additional chapter files, simply merge the new `InfraWorks Drainage Essentials` folder into the old one.

Configure Your Units

When you begin using InfraWorks, you will need to configure it for either Imperial or metric units. Here are the steps:

1. Launch InfraWorks 360.
2. On the start page, click Application Options.
3. In the Application Options dialog, click Unit Configuration.
4. For Default Units, select either Imperial or Metric based on the units you prefer to work in.

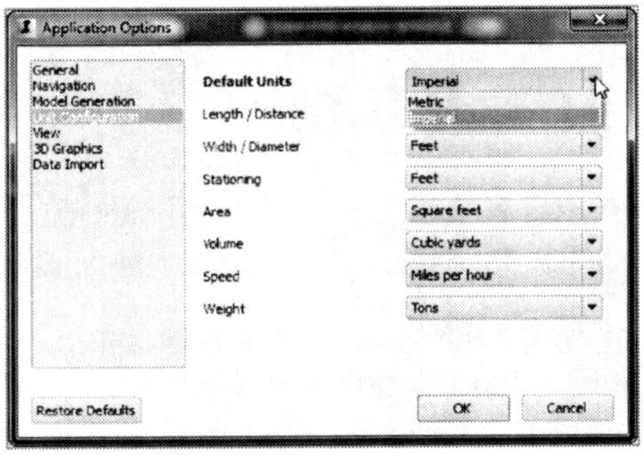

Know Which Units to Use

As you work through the book, you will see dimensions provided twice: once in Imperial units and then again as metric units in parentheses. The following is an example step from an exercise:

9. Click the height gizmo and enter a value of 150 (45) in the tooltip. Press Enter. Press Esc to clear the selection of the building.

In this example, you would enter 150 if you are working in imperial units and 45 if you are working in metric units. Note that 45 meters is not the exact metric equivalent of 150 feet but a practical approximation.

Check for Updates

Finally, be sure to check the book's website, `www.sybex.com/go/drainageessentials`, for any updates to this book should the need arise. You can also contact me directly by email at `InfraWorksEssentials@gmail.com`.

The Essentials Series

The Essentials series from Sybex provides outstanding instruction for readers who are just beginning to develop their professional skills. Every Essentials book includes skill-based instruction with chapters organized around projects rather than abstract concepts or subjects, plus digital files (via download) so you can work through the project tutorials.

What Is Covered in This Book

Autodesk Drainage Design for InfraWorks 360 Essentials is organized to provide you with the knowledge needed to master the basics of the Drainage Design module.

Chapter 1: Getting Started Familiarizes you with the concept of Drainage Design for InfraWorks 360 and gives you a detailed account of all the tools available through its user interface

Chapter 2: Designing and Analyzing Drainage Demonstrates how to create, modify, and analyze watersheds, culverts, and drainage networks utilizing a number of tools

How to Contact the Author

I welcome feedback from you about this book or about books you'd like to see from me in the future. You can reach me by writing to `InfraWorksEssentials@gmail.com`. For more information about my work, please visit my blog at `http://erichappell.blogspot.com`.

Sybex strives to keep you supplied with the latest tools and information you need for your work. Please check its website at `www.sybex.com/go/drainageessentials`, where we'll post additional content and updates that supplement this book if the need arises.

CHAPTER 1

Getting Started

Autodesk® Drainage Design for InfraWorks 360™ is an add-on module that runs in Autodesk® InfraWorks 360™. It provides additional functionality for performing hydrologic and hydraulic design and analysis for land development projects. It is an excellent tool for performing stormwater design and analysis in the early stages of a project.

In this chapter, you'll learn to

- **Identify and understand the capabilities of Drainage Design for InfraWorks 360**
- **Understand and navigate the additions to the InfraWorks user Interface**

Understanding the Capabilities of Drainage Design for InfraWorks 360

Drainage Design for InfraWorks 360 extends the capabilities of InfraWorks 360 by offering advanced design tools for hydrologic and hydraulic analysis and design. To gain an understanding of how these advanced capabilities are integrated into the software, you'll look at several areas, including watershed analysis, culvert design, pavement drainage design, and quantities.

Watershed Analysis

Before designing infrastructure that will serve the drainage needs of your site, you must first calculate the flows that the infrastructure will be required to handle. This is determined through hydrologic analysis, and the watershed tools in Drainage Design for InfraWorks are built for this purpose. You will study the capabilities of these tools in two major areas: delineating watersheds and calculating flows.

Delineating Watersheds

When analyzing the hydrology of a project, one of the first tasks that must be performed is delineating watersheds: areas that drain to specific points of interest. In Figure 1.1 you see a watershed shaded in blue, which represents the area that drains to a point beneath a bridge. In other words, when there is a rain storm, every drop of rain that hits the ground within that blue area (and is not absorbed into the ground) will flow beneath the bridge. In many cases, designers are most interested in analyzing the amount of stormwater flow that enters a structure, such as a culvert or inlet. Many methods are available for determining that flow, but most, if not all of them, require that the watershed for that structure has been determined.

FIGURE 1.1 The blue shaded area is a watershed contributing flow under the bridge shown near the center of the image.

The process for delineating a watershed requires the analysis of the terrain to determine how the varying slopes of your site direct runoff toward and away from a given point of interest. This requires some serious computations, and that's where the cloud computing power of InfraWorks 360 plays a key role. Drainage Design for InfraWorks 360 has the capability to harness the power of the cloud to compute and delineate drainage areas for individual points or entire stretches of road (see Figure 1.2).

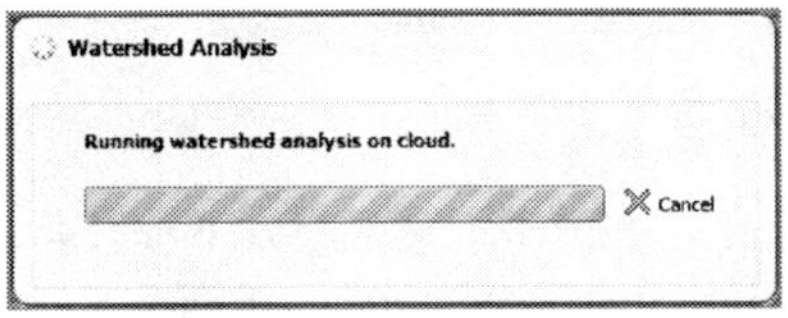

FIGURE 1.2 **InfraWorks is leveraging the cloud to perform a watershed analysis.**

Calculating Flow

Once a watershed is delineated, the next step in the process is to calculate the flow to the point of interest. Across the world, many different methods are used to calculate this flow, but there are common themes among all of them. See the sidebar "One Method: The Rational Method" to get an idea how one of these methods works. InfraWorks enables you to choose from a number of methods to determine how to calculate the predicted flow for a given point of interest. Where applicable, you can also supply important information that is used in the calculations. This is all done through the Watershed asset card (see Figure 1.3).

FIGURE 1.3 **The Watershed asset card is being used to calculate the flow for the watershed shown to the left of it.**

One Method: The Rational Method

If you are new to the concepts of hydrologic analysis, the Rational method is a good place to start. The Rational method uses a simple equation to calculate the predicted flow to a given point based on the terrain, land cover, and historical rainfall data. The formula for the Rational method is $Q = CIA$.

- Q is the predicted flow measured in units of volume per units of time. For example, cubic feet per second (CFS) is often used for Imperial units, and cubic meters per second (CMS) is used for metric units.
- C is the runoff factor, which ranges from 0 to 1. Nearly impervious surfaces such as asphalt or concrete are represented with high C values in the 0.7–0.9 range. Densely forested areas are much lower, in the 0.1–0.3 range. And everything else falls somewhere in between. The C factor represents the amount of rainfall that is *not* absorbed into the ground and therefore allowed to flow over it. For example, an area that is paved may have a C factor of 0.85, which is a way of saying that 15 percent of the rainfall will be absorbed into the ground and the remaining 85 percent will stay on top of the ground and become runoff.
- I in the Rational equation stands for intensity and indicates the expected severity of the rainfall. Time of concentration also factors into determining the I value for the Rational method. The time of concentration is the amount of time that it takes for every bit of a watershed to contribute flow to the point of interest (see the sidebar "Concentrate to Understand Time of Concentration").
- A is simply the area of the watershed contributing to the point of interest.

So, if you simply think about the equation $Q = CIA$, you realize that watersheds that are more impervious (higher C), see more intense rainfall (higher I), and cover larger areas (higher A) will generate higher flows (Q). Makes sense!

Concentrate to Understand Time of Concentration

What is time of concentration, and why should you care about it? When analyzing watersheds, you are often most concerned with peak flow—the highest flow that will occur during a storm event. If you were to make a graph of this flow (called a *hydrograph*), it typically has a bell shape, with the flow starting small, increasing to a point, and diminishing to zero. At first you might think that happens because the rain starts out light and then gets heavier and finally subsides, but that's not the reason. Even if you could turn the rain on with a switch and have it start and end with the same intensity, you would still see this bell curve representing the flow. What gives?

The rain that falls must travel a certain distance to reach the point of interest: the outlet of the watershed. A few seconds into the rain event, only the raindrops that have fallen closest to the outlet have had enough time to get there, so the flow is small. After a minute, rain drops that fell, say, 100 feet away are now reaching the outlet *plus* all the ones that are continuing to fall between 0 and 100 feet away, so the flow increases. After five minutes, raindrops are arriving that fell, say, 500 feet away *plus* all of the ones that are continuing to fall between 0 and 500 feet away, so the flow increases. Eventually, the last spot in the watershed, a corner that is maybe 752 feet away, has finally provided a raindrop to the outlet, and perhaps it took 7.36 minutes for it to get there. Now, every square inch of the watershed is contributing flow to the outlet, and the flow has reached its peak. The time of concentration is 7.36 minutes.

Make sense? If so, take a deep breath and get ready for a little more. That last raindrop is not necessarily the furthest away—technically it is simply the one that took the longest to get there. It is possible that a raindrop that is far away traveled down a path quickly because there was little resistance (perhaps a curb or steep concrete channel), whereas a closer one had to travel across a forest floor that was relatively flat. So, figuring out the time of concentration involves more than just measuring how far it is from the outlet to the furthest spot in the watershed. There's more, but I'll leave it at that.

Culvert Design

As you probably already know, road design is a big part of InfraWorks—so much so that there is a module dedicated to it named Autodesk® Roadway Design for InfraWorks 360™. Whenever roads are built across the landscape, there are always drainage concerns to consider, and one of the primary ones is allowing drainage ditches, streams, and rivers to pass beneath the road. Bridges are required for larger waterways (by the way, there is a module for bridges too: Autodesk® Bridge Design for InfraWorks 360™), but for small-flow applications, culverts are the solution. Drainage Design for InfraWorks 360 contains a powerful set of culvert design tools that enable you to quickly create, design, and adjust culverts (see Figure 1.4).

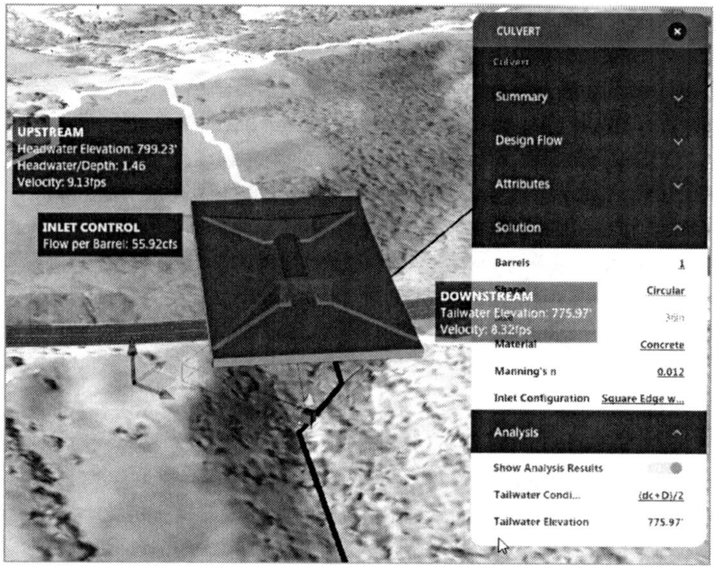

FIGURE 1.4 The configuration of a culvert where a design road crosses a stream

Automated Culvert Design

When you use the Create Watersheds command to delineate watersheds in your model, another product of that command is a series of stream networks. If any stream in a given stream network crosses beneath a road, with a single command you can have InfraWorks automatically create culverts at those locations.

And not only will the culverts be created at the proper location, they will be properly sized based on the flows calculated for the associated watersheds. Better yet, when you make changes to the road, such as adding lanes or adjusting the alignment, the culverts will automatically update to accommodate. Culverts understand that they are the products of the interactions between the roads, watersheds, and stream crossings, and they will respond to changes in any of those components.

If you do not want to perform a watershed analysis, you can still ask InfraWorks to create culverts automatically at places where rivers cross beneath a road. In this case, the culverts will not "know" their design flows and will therefore not be sized, but they will respond to changes in the design of the road such as changing the number of lanes, moving the road, or changing where the road ties to the existing terrain. Figure 1.5 shows a culvert that has been placed in this manner.

FIGURE 1.5 A culvert that has been placed where a river crosses beneath a road

Manual Culvert Placement

You can also place a culvert in the model manually by launching the Culvert command and simply picking two points to indicate the start and end. The inverts of the culvert will automatically match the terrain by default, but you can change those elevations using familiar techniques such as dragging gizmos or using the Properties panel. This approach, although not tied to any other design components, gives you complete control over the placement of a culvert. The culvert in Figure 1.6 was placed manually with no stream or water area present.

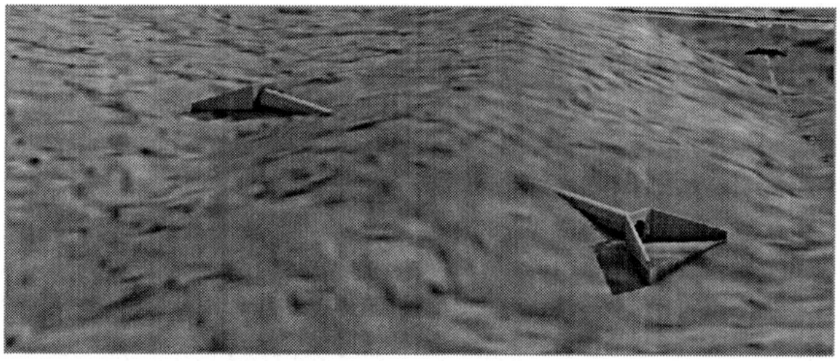

FIGURE 1.6 A manually placed culvert

Culvert Content

As you design culverts using Drainage Design for InfraWorks 360, you'll be impressed by the number of choices you have for culvert shapes, number of barrels, and end configurations. With all these choices, you can perform detailed designs that conform to your project conditions as well as configurations that match local construction practices and available materials. Figure 1.7 shows a user choosing the wingwall configuration for a culvert.

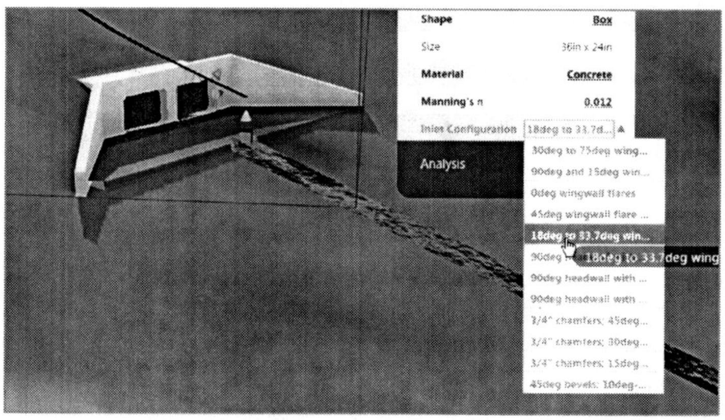

FIGURE 1.7 Choosing a wingwall configuration

Culvert Analysis

Once you have placed and sized a culvert, InfraWorks will perform a basic analysis and provide results to you in two ways. The first is a visual analysis showing you a 3D graphic of the hydraulic grade line along with key values relating to elevations, depths, flows, velocities, and so on (see Figure 1.8). The second way that InfraWorks provides results is by giving you the ability to export the results to a file that can be imported into another product for further analysis.

FIGURE 1.8 Culvert analysis

Drainage Networks

Another important drainage design consideration is the collection and conveyance of runoff on the road surface itself. Drainage Design for InfraWorks 360 provides the ability to create drainage networks for this purpose, either automatically or manually. In much the same way that InfraWorks will create, position, and design culverts for you, it will also place inlets, manholes, and pipes along a design road automatically based on design rules and common engineering assumptions. Figure 1.9 uses a transparent road style to show a portion of a drainage network that was automatically created in this manner.

FIGURE 1.9 A drainage network created automatically for a road

Once in place, these drainage network components can be analyzed and modified (see Figure 1.10) so that the design can be adjusted to meet the needs of the project. Drainage network components can also be placed manually to add more detail to an automatically created network or to simply create a network from scratch, having full control of the placement of the components.

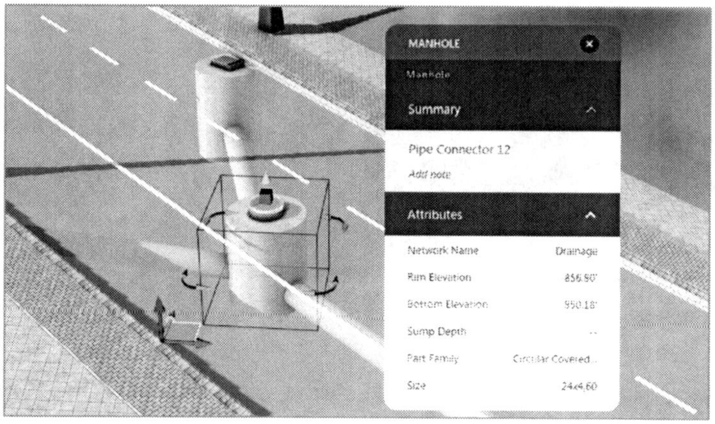

FIGURE 1.10 Analyzing and modifying a drainage network component

Quantities

Another useful task that Drainage Design for Infraworks 360 will perform is tabulating drainage quantities for a given road. This comes in handy when performing cost analysis for a given road by providing quantities for culverts as well as drainage network components (see Figure 1.11).

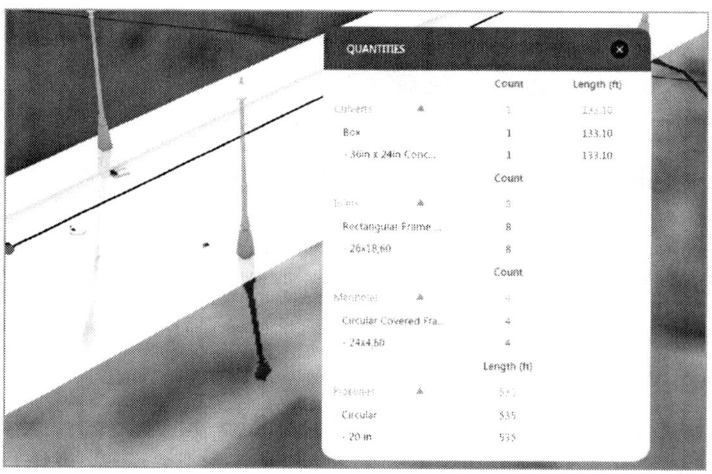

FIGURE 1.11 Drainage quantities for a road

The Drainage Design User Interface

As you might guess, Drainage Design for InfraWorks 360 adds some new user interface components to InfraWorks 360. These include additions to the Intelligent Tools, asset cards, and right-click menus.

Drainage Design Intelligent Tools

With Drainage Design for InfraWorks 360 enabled, you will see a bluish green pipe icon on the main toolbar that I'll call the Drainage Design icon. This icon expands to reveal the Drainage Design toolbar, which contains an Analyze icon, a Design icon, and a Review icon (see Figure 1.12). Each one of these icons opens a toolbar along the left side of the screen. I'll refer to these toolbars as the Analyze, Design, and Review toolbars. As with other InfraWorks modules, these toolbars contain a mixture of basic InfraWorks commands along with commands that are exclusive to Drainage Design for InfraWorks 360.

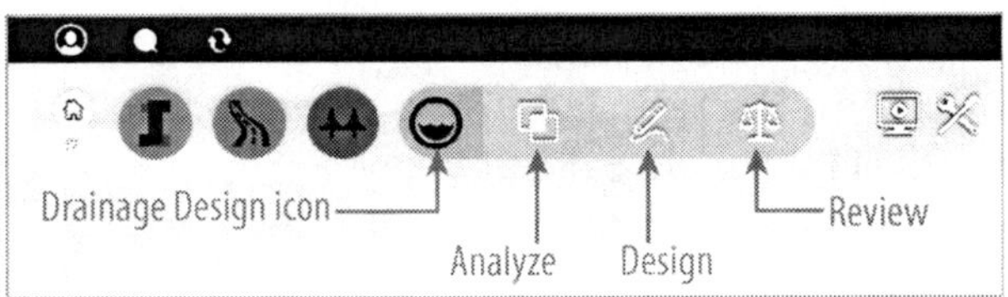

FIGURE 1.12 The Drainage Design toolbar

Analyze Toolbar

The Analyze toolbar contains four tools; two are part of the core InfraWorks tools, and two are from the Drainage Design tools.

Terrain Themes Terrain Themes is a basic InfraWorks command. It opens the Terrain Themes panel where you can configure a theme to color-code a terrain based on elevation, slope, or aspect (direction in which hillsides face).

Create Watersheds This command uses the cloud capabilities of InfraWorks 360 to analyze the terrain of the model and identify watersheds and streams. It creates new watershed and stream objects in the model that you can use for further design and analysis.

Surface Layers Surface Layers is a basic InfraWorks command. Surface layers are the parts of the model that rest on the ground. This icon opens a panel that lets you manage the visibility of these objects.

Quantities This prompts you to select a design road, and when you do, it opens the Quantities asset card, which reports culvert and drainage network quantities to you.

The Design Toolbar

The Design toolbar contains eight tools; three are part of the core InfraWorks set of tools, and five are from the Drainage Design tools.

Coverages Coverages is a basic InfraWorks command. You can think of a coverage as an area of land that is covered with something. That can be grass, pavement, concrete, sand, or just about anything. Coverages can also be used to shape the land that they cover.

Culverts This will launch the manual culvert tool and prompt you to pick two points defining either end of a culvert.

Points Of Interest Points Of Interest is a basic InfraWorks command. You use this command to create points of interest to call out important locations in your model. Any 3D model can be used as a marker for a point of interest.

Rivers Rivers is a basic InfraWorks command. Use this command to draw rivers, creeks, canals, and other linear waterways by clicking points along a path.

Water Areas Water Areas is a basic InfraWorks command. You use this command to draw lakes, ponds, and other water features defined by closed shapes.

Pavement Drainage You use this command to create inlets, manholes, and pipelines that represent drainage networks in your model.

Create Watersheds This command will use the cloud capabilities of InfraWorks 360 to analyze the terrain of the model and identify watersheds and streams. It will create new watershed and stream objects in the model that you can use for further design and analysis.

Style Palette Style Palette is a basic InfraWorks command. The appearance of objects in InfraWorks is controlled by styles. Every InfraWorks model has a library of styles that you can pull from and apply to objects to quickly change their appearance. The Style Palette command presents that library to you in an organized fashion. You can also create your own styles and add them to the Style Palette.

The Review Toolbar

The Review toolbar contains six tools; five are part of the core InfraWorks set of tools, and one is from the Drainage Design tools.

Surface Layers Surface Layers is a basic InfraWorks command. Surface layers are the parts of the model that rest on the ground. This icon opens a panel that lets you manage the visibility of these objects.

Profile View This tool opens the Profile View panel, which has the primary purpose of viewing and editing road designs. However, you can also see markers indicating drainage components.

Surface Opacity This is a basic InfraWorks feature. It is a handy toggle that switches your terrain surface from see-through to opaque. It is great for working on underground features such as culverts or drainage networks.

Point-To-Point Distance This is a basic InfraWorks feature. You use it to measure a straight-line distance by picking two points in the model.

Path Distance This is a basic InfraWorks feature. You use it to measure the distance along a path by picking multiple points in the model.

Terrain Statistics This is a basic InfraWorks feature. You use it to get information about an area of terrain such as area, volume, and other data. You can also get data about how a coverage is modifying existing terrain such as cut, fill, and other design information.

Drainage Design Asset Cards

Seven new asset cards come along with Drainage Design for InfraWorks 360: Create Watersheds, Watershed, Culvert, Quantities, and three Drainage Network asset cards. You will study these in greater detail in Chapter 2.

Create Watersheds The Create Watershed asset card is used to configure key settings before submitting your watershed analysis for processing in the cloud. It also displays information and provides links relating to cloud credits.

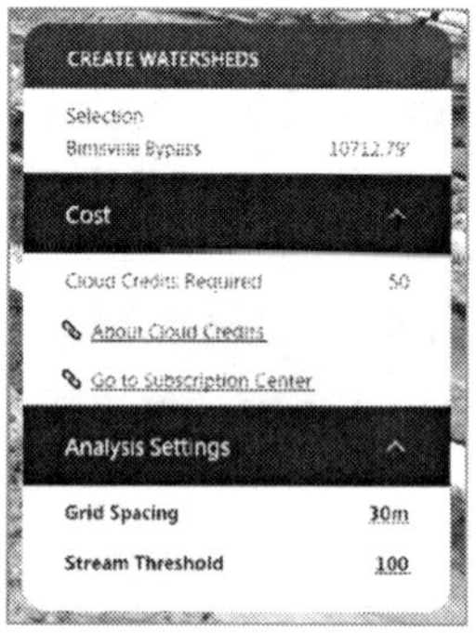

Watershed The Watershed asset card is used for reporting information to you as well as configuring the watershed for hydrologic analysis. From here you can view information about the area, elevations, and slopes of the watershed as well as choose a calculation method, assign key values, and ultimately calculate flows.

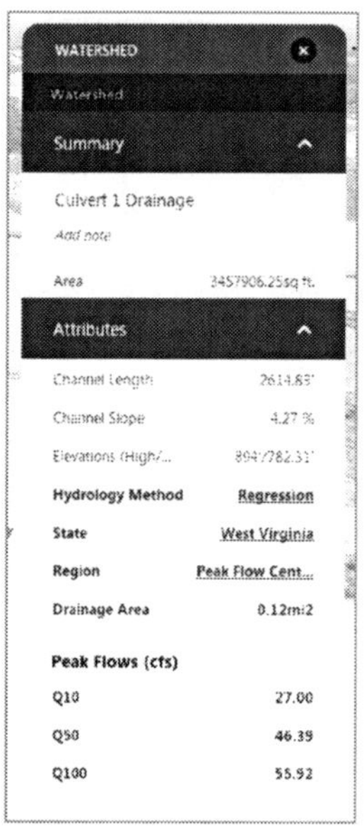

Culvert The Culvert asset card provides you with full design capabilities for a culvert while providing simultaneous analysis. Here you can choose the culvert shape, number of barrels, inlet configuration, and other important design features. You can also define the tailwater condition and elevation and toggle the graphical view of the culvert analysis.

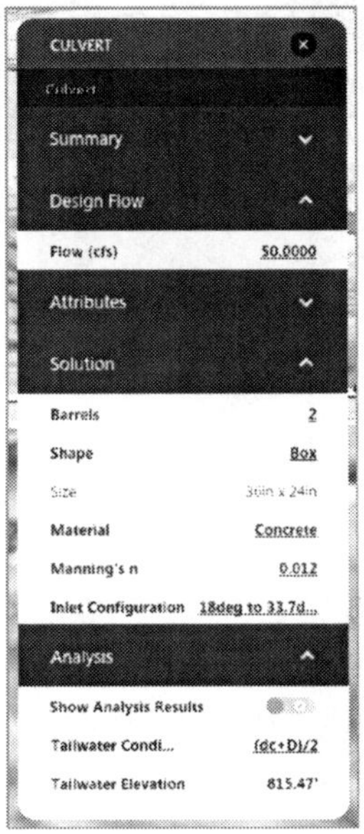

Quantities The Quantities asset card reports quantities of culverts, inlets, manholes, and pipelines for a design road that you choose. No information can be edited on this asset card.

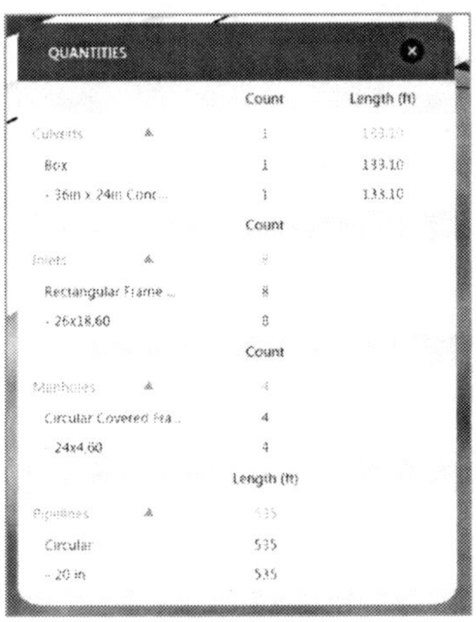

Drainage Network The three drainage network asset cards, Inlet, Manhole, and Pipeline, report information about the drainage network component you have selected. Other than providing a name for the component and adding a note, no information can be changed on these asset cards.

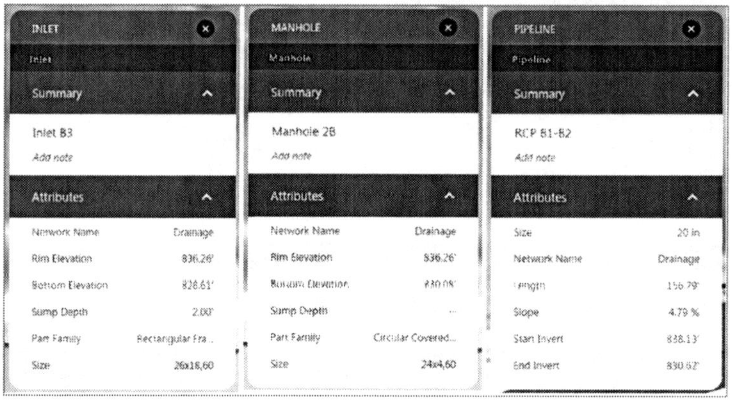

Right-Click Commands

With Drainage Design for InfraWorks 360 enabled, you will find some additional commands in several right-click menus. These additions are covered next.

Roads

When you right-click a road while in edit mode, you will see three commands that relate to drainage design (see Figure 1.13).

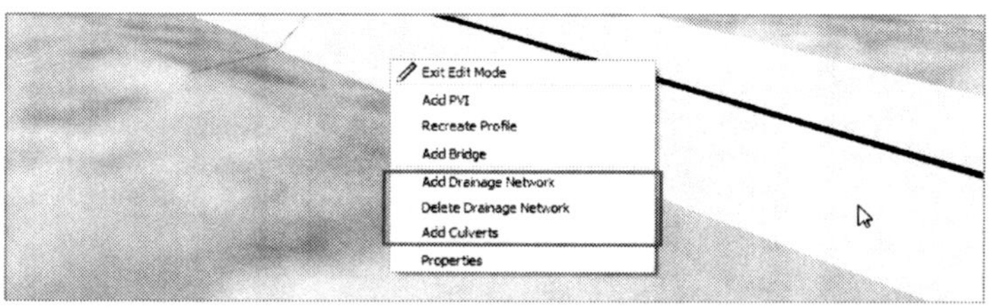

FIGURE 1.13 The additional drainage commands are within the red box.

Add Drainage Network This command automatically adds inlets, manholes, and pipelines along the design road based on design rules and engineering assumptions. You can edit the network once it is created. This command appears only if the road has curbs.

Delete Drainage Network This command deletes all inlets, manholes, or pipelines that are part of a drainage network associated with the design road.

Add Culverts This command analyzes the road and inserts a culvert wherever the road crosses a water area or stream. The culvert will remain associated with the road and the water feature that it crosses. It will update automatically as edits are made.

Culverts

When you right-click a culvert while in edit mode, you will see a shortcut menu that contains four commands (see Figure 1.14).

FIGURE 1.14 Right-click menu for culverts

Exit Edit Mode As you may have seen with other InfraWorks objects, this turns off edit mode, causing the gizmos to disappear and the editing functions to be disabled.

Analyze Culvert This command reassesses aspects that affect the design of the culvert such as flow or road geometry. Most of the time, the culvert updates on its own, but this command can be used to manually trigger an update.

Export Analysis to CSV Just like the name implies, this command exports the culvert analysis results to a comma-delimited text file with a `.csv` extension. The information in this file can be placed in a report or utilized in another program for further analysis.

Properties This is the standard Properties command that opens the Properties panel and displays information about the culvert.

Pavement Drainage Layout

When you launch the Pavement Drainage command, you can right-click and choose which components you would like to be inserted as you click points in the model. The choices are Inlets And Pipelines, Manholes And Pipelines, Inlets Only, and Manholes Only (see Figure 1.15).

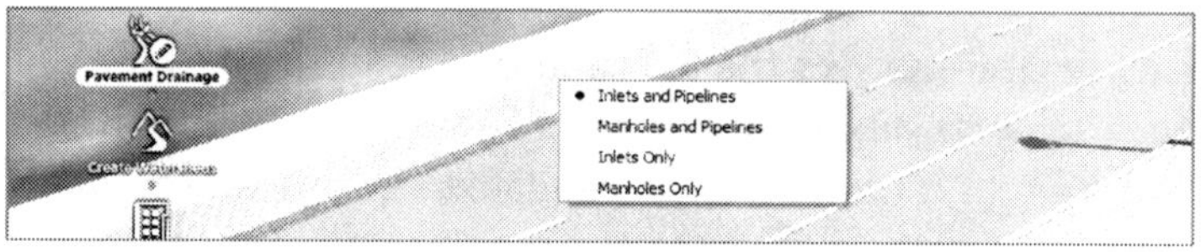

FIGURE 1.15 The right-click menu available while laying out pavement drainage

Drainage Properties

With Drainage Design for InfraWorks 360 enabled, you will experience the addition of several new InfraWorks objects such as culverts, watersheds, and streams. These objects will have their own unique properties in addition to the "generic" properties that all InfraWorks objects have. Drainage network components are still classified as pipelines and pipeline connectors, but they do possess some differences when compared with those objects created with basic InfraWorks tools. Figure 1.16 shows the properties of a watershed, which you can see has unique values such as those shown in the Watershed section.

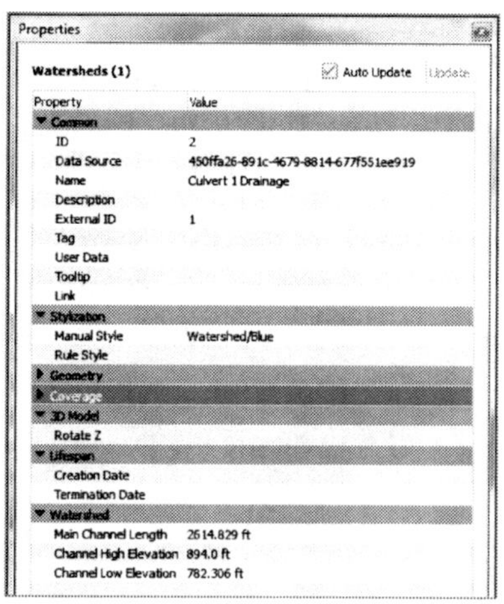

FIGURE 1.16 The properties of a watershed

Exercise 1.1: Explore Drainage Design for InfraWorks 360

To begin this exercise, go to the book's web page at `www.sybex.com/go/drainageessentials` and download the files for Chapter 1. Unzip the files to the correct location on your hard drive according to the instructions in the introduction.

1. Launch InfraWorks 360.
2. On the Start Page, click Open.
3. Browse to `C:\InfraWorks Drainage Essentials\Chapter 01\` and select `Ch01 Bimsville Drainage.sqlite`. Click Open.

   You are looking at a model of the Bimsville Bypass project. If you have worked through the book *Autodesk® InfraWorks and InfraWorks 360™ Essentials*, then you are quite familiar with this project. In this version of the model, only a portion of the bypass has been created, and only the roads for the industrial park are in the model.
4. If the Ex_1_1 proposal is not already current, click the Proposals drop-down list on the Utility Bar and select it, as shown in Figure 1.17.

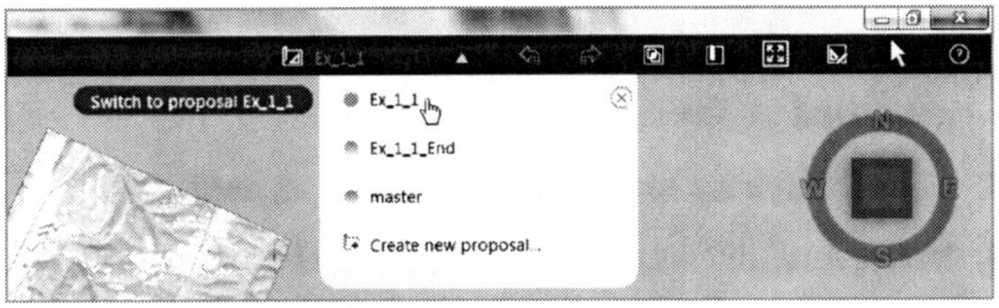

FIGURE 1.17 Selecting the correct proposal

5. Click the Bookmarks icon on the Utility Bar; then click Watershed 1 to restore that bookmark.

   You are looking at a location where a watershed is shown that feeds a small stream that crosses the Bimsville Bypass. This watershed was generated using the Create Watersheds command.

6. Click within the shaded area representing the watershed. The Watershed asset card should open, as shown in Figure 1.18.

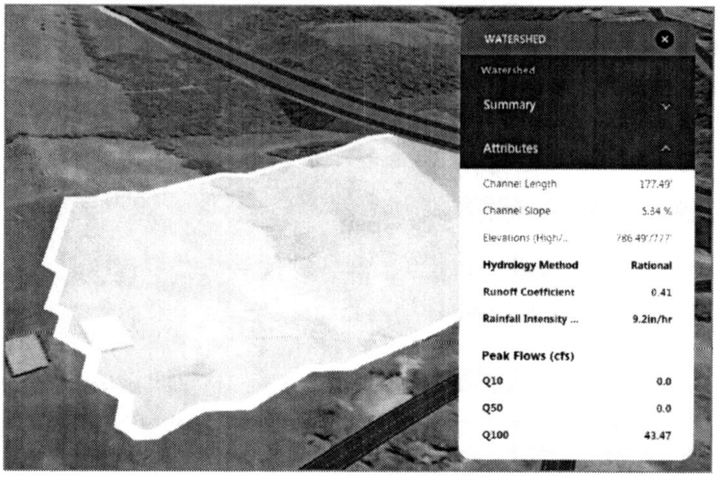

FIGURE 1.18 A watershed along with the Watershed asset card

Scan some of the information shown in this asset card; note that a peak flow has been calculated for the Q100 value.

7. Press Esc to clear the selection and close the asset card. Restore the bookmark named Culvert 1.

   You are looking at one end of a culvert created beneath the Bimsville Bypass. This culvert was created using the Add Culverts command on the road right-click menu.

8. Click the culvert to open the Culvert asset card.

9. In the Analysis section of the Culvert asset card, click the switch next to Show Analysis Results to turn it on. If it is already on, skip to the next step.

10. Restore the bookmark named Culvert 1 Analysis. You can now see the graphical representation of the culvert analysis hovering above the actual culvert (see Figure 1.19).

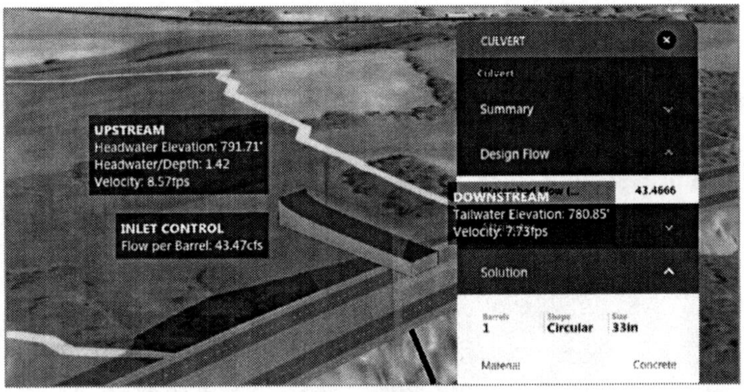

FIGURE 1.19 **The graphical analysis of a culvert along with its asset card**

11. Press Esc to clear the selection of the culvert, close the analysis, and close the asset card. Restore the bookmark named Pavement Drainage.

    You are now viewing two inlets and a manhole that have been placed on a two-lane design road. If you are familiar with any of the other InfraWorks Essentials books, you may recognize this road as the main road of the industrial park. The buildings, coverages, and other aspects of the industrial park are not included in this model.

12. Restore the bookmark named Pavement Drainage Underground.

    From an underground vantage point, you are now viewing the same structures along with pipelines and other structures located elsewhere along the road (see Figure 1.20).

If you are unable to navigate underground, make sure that the Lock Mouse Navigation Above Ground option is turned off in the Application Options dialog, Navigation view.

FIGURE 1.20 **An underground view of a drainage network**

13. Restore the bookmark named Pavement Drainage. Click the road surface to select it and right-click.

14. If the command at the top of the right-click menu says Exit Edit Mode, do nothing. If instead you see Edit at the top of the menu, click Edit and right-click again.

    Figure 1.21 shows the drainage related commands that are shown in this menu.

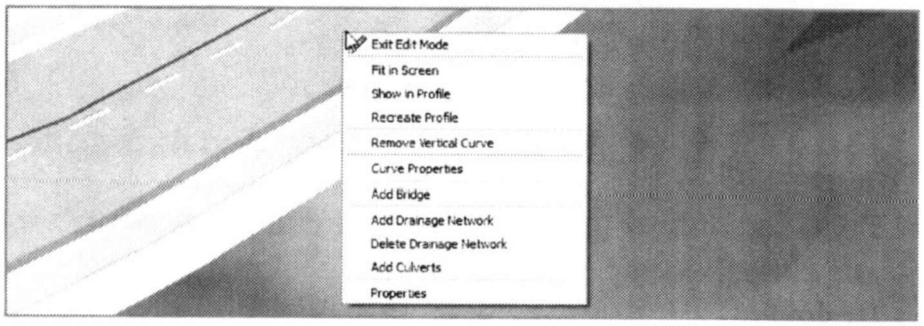

FIGURE 1.21 The right-click menu for a design road that contains several drainage-related commands

15. Click Delete Drainage Network.

    The structures and pipelines will disappear from your current view. If you zoom out, you will notice that all of the structures and pipelines have been deleted for the entire road.

    You can view the results of successfully completing this exercise by selecting the Ex_1_1_End proposal.

Now You Know

Now that you have completed this chapter, you have an understanding of the capabilities of Drainage Design for InfraWorks 360. You understand that it runs in InfraWorks 360 and extends its functionality by providing tools for watershed analysis, culvert design, and drainage network design. You also have an understanding of what Drainage Design for InfraWorks 360 adds to the InfraWorks user interface. This includes several new intelligent tools, asset cards, and right-click commands. This also includes some additional properties that you can view and modify in the Properties panel.

After completing this chapter, you are comfortable navigating within Drainage Design for InfraWorks 360, and you're ready to begin using it to perform some serious drainage design.

CHAPTER 2

Designing and Analyzing Drainage

Now that you've familiarized yourself with Autodesk® Drainage Design for InfraWorks 360™ and are aware of its capabilities, you are ready to begin performing your own analyses and designs. As you'll learn in this chapter, the technology that enables you to design and analyze drainage in InfraWorks is quite powerful but also relatively simple to use. In this chapter, you'll look at several topics that relate to drainage design and analysis.

In this chapter, you'll learn to

- **Create watersheds**
- **Analyze watersheds**
- **Design culverts**
- **Design drainage networks**
- **Quantify your drainage design**

Working with Watersheds

In this section you will study the use of Autodesk® InfraWorks™ to create and analyze watersheds. I'll start with a discussion of the "science" of watersheds, what they are, and why they're important when performing drainage analysis and design. Then I'll show how to create watersheds and finish up with how to analyze watersheds to determine predicted peak flows.

Understanding Watersheds

A watershed is an area that drains to a specific location such as a culvert entrance, inlet opening, bridge opening, stream, river, pond, or lake (see Figure 2.1).

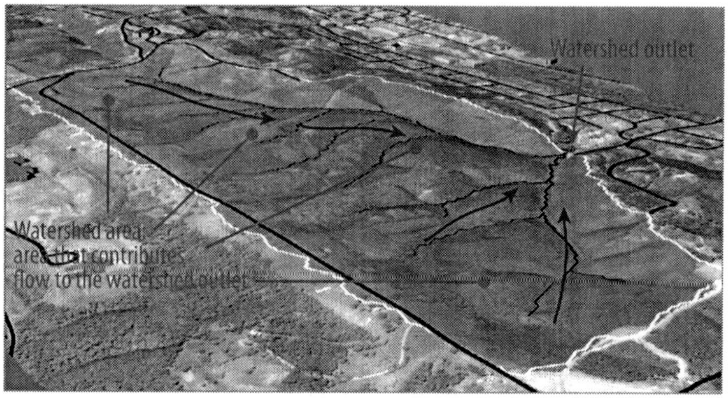

FIGURE 2.1 An example of a watershed

Several important characteristics of the watershed determine the amount of flow that will leave the watershed and enter the location of interest. Of these characteristics, the one that is easiest to understand is area. The larger the area that a watershed covers is, the more flow it will contribute to the location of interest. The second characteristic is the amount of water that will flow across the ground surface vs. being absorbed. Many factors impact this characteristic such as the soil type, the topography, and the nature of the ground surface: pavement vs. grass vs. forest, for example. Since these characteristics can vary across a given watershed, either assumptions must be made or a detailed analysis must be performed to predict how readily the watershed will absorb water. This is usually expressed in some sort of runoff factor.

The shape and area of a watershed are determined by the topography and possibly man-made structures such as walls, dams, channels, and so on. Generally speaking, the highest points in a watershed form the outer boundary, and the lowest point is the location where runoff exits the watershed. In addition, the direction that the land slopes has a direct effect on the direction that runoff flows. To delineate a watershed, the terrain must be analyzed to find these ridge lines and low points and determine the directions of slopes.

Creating Watersheds

As you might guess, the terrain analysis described in the previous section requires rather intense analysis and calculation, the type that is done with ample computing power. Since most individuals do not possess computers with this amount of power, the watershed functionality of InfraWorks looks to the cloud to perform its tasks. To start the process, you click the Create Watersheds icon on either the Analysis toolbar (see Figure 2.2) or the Design toolbar of the Drainage Design tools.

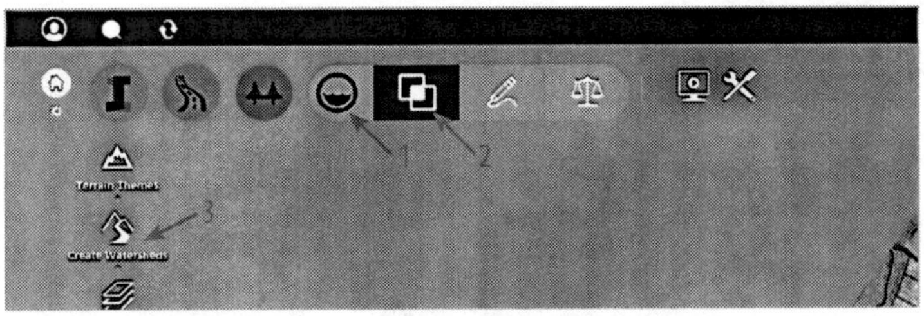

FIGURE 2.2 **You can click the icons in the order shown to create one or more watersheds.**

Once you have launched the tool, you can either click a design road or click a point in the model that is not on a design road. If you click a design road, you can either press Enter to analyze the entire road or pick two points along the road to confine the analysis to a region of the road (see Figure 2.3). If you pick a point on the model that is not on a road, you can simply begin the analysis by pressing Enter.

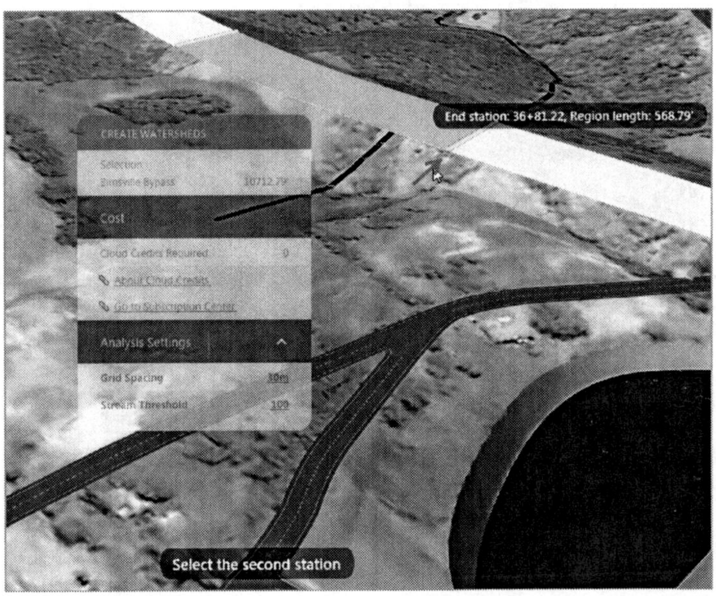

FIGURE 2.3 **Defining a region on a design road for watershed analysis**

When you launch the Create Watersheds command, the Create Watersheds asset card will open (see Figure 2.4). You can modify the applicable values on this asset card at any time before pressing Enter to start the analysis process. The first two sections of the asset card simply report what you have selected and what the cost will be in cloud credits. Neither of the values in these two sections can be modified. The Analysis Settings section contains two values that you will want to address. The first is Grid Spacing, and the second is Stream Threshold.

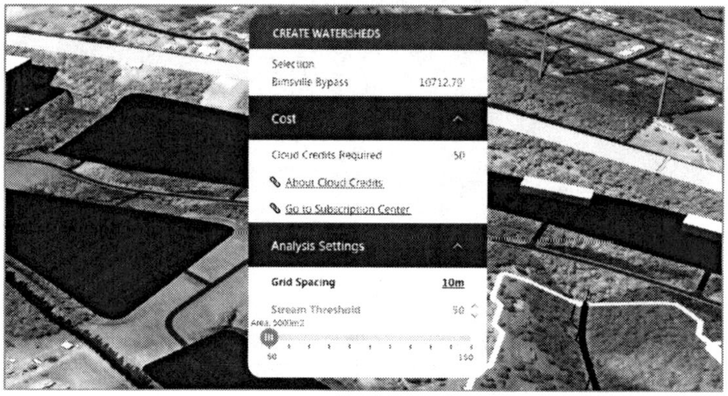

FIGURE 2.4 **The Create Watersheds asset card**

Grid Spacing Your terrain will be analyzed by first subdividing it into grid squares that have a constant slope. These grid squares become the "units" of the analysis. The limits of the watershed and the paths of the streams will be drawn along the edges of these grid squares. The smaller you make these squares, the more accurate your analysis will be—but at the expense of computing time and possibly model performance. Therefore, choose a value that is appropriate for the size of your project and the accuracy that is required. The value ranges from 10 meters to 50 meters. (There is no Imperial equivalent for this.)

Stream Threshold This value sets the number of grid squares used to define streams. For example, a Stream Threshold of 50 means that a stream is created once there are 50 grid squares contributing flow to its starting point. As you set the value for Stream Threshold, you will see an area value reported at the left end of the slider. So, for example, if Grid Spacing is set to 10 meters and Stream Threshold is set to 50, then the area will be 5,000 square meters, the equivalent of 50 10×10 squares. In other words, a smaller value here will generally produce longer, more numerous streams, and a larger one will produce fewer streams that are shorter. The value for Stream Threshold can range from 50 to 150.

Cloud Credits?

Certain functions in InfraWorks 360 require the use of cloud credits, and the Create Watersheds command is one of them. Cloud credits must be purchased by you or your company and allocated by an account administrator. If you are the sole owner of your software, then that is you. If you work for a company or organization, you may need to request to have cloud credits allocated to your account.

Once you have selected your point, road, or road region and you have configured the Create Watersheds asset card to your requirements, you can press Enter to begin the analysis process. InfraWorks will prepare the necessary data, upload it to the cloud, process the data, and finally generate your results in the model in the form of one or more watersheds, stream networks, and points of interest indicating outlet points or points where streams cross design roads. You will see these items listed on the Data Sources panel (see Figure 2.5) where they can be managed like any other data source you have learned about. Also, at the end of the process, InfraWorks will report how many cloud credits were deducted from your account as a result of the analysis.

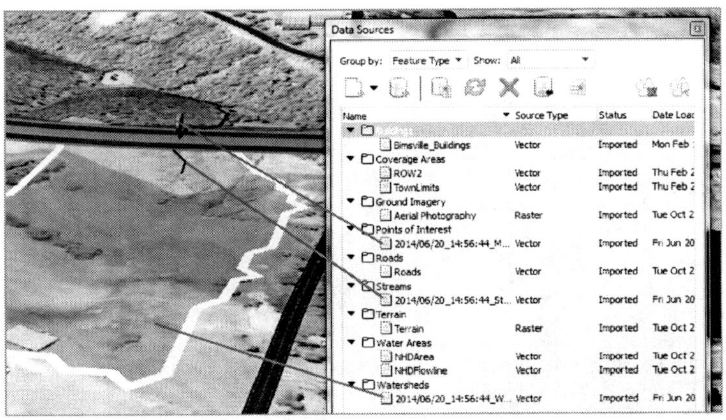

FIGURE 2.5 A watershed, stream, and point of interest shown in the model as well as in the Data Sources panel

Exercise 2.1: Create a Point Watershed

In this exercise you will create a watershed that drains to the location of a future culvert. You will indicate the location of the culvert entrance and allow InfraWorks to calculate the watershed limits.

To begin this exercise, go to the book's web page at www.sybex.com/go/drainageessentials and download the files for Chapter 2. Unzip the files to the correct location on your hard drive according to the instructions in the introduction.

1. Launch InfraWorks 360.
2. On the Start Page, click Open.
3. Browse to C:\InfraWorks Drainage Essentials\Chapter 02\ and select Ch02 Bimsville Drainage.sqlite. Click Open.
4. If the Ex_2_1 proposal is not already current, select it from the Proposals drop-down menu.
5. Click the Bookmarks icon on the Utility Bar; then click Point Watershed to restore that bookmark.

   You are looking at a location where the embankment of the industrial park road is obstructing the natural flow of water in a small valley (see Figure 2.6). As a designer you may want to place a culvert entrance at this location to allow the flow to continue beneath the road. To design that culvert, you will need to calculate the flow to its entry point. In the upcoming steps, you will create a watershed for this location so that this flow can eventually be calculated.

FIGURE 2.6 A design road is obstructing the flow of water.

6. Click the Drainage icon on the main toolbar to expand the Drainage toolbar.
7. Click the Design icon on the Drainage toolbar.
8. Click Create Watersheds on the Design toolbar. If a dialog opens informing you that an Internet connection and InfraWorks 360 account are required, click OK.

   The Create Watersheds asset card will open, and a prompt will appear at the bottom of the screen instructing you to click an outlet point or select a design road.
9. Click a point in the model at the low point where the road embankment meets the bottom of the valley. The location of this point would be near the tip of the blue arrow in Figure 2.6.

   A blue point of interest will be inserted at the location you picked, as shown in Figure 2.7.

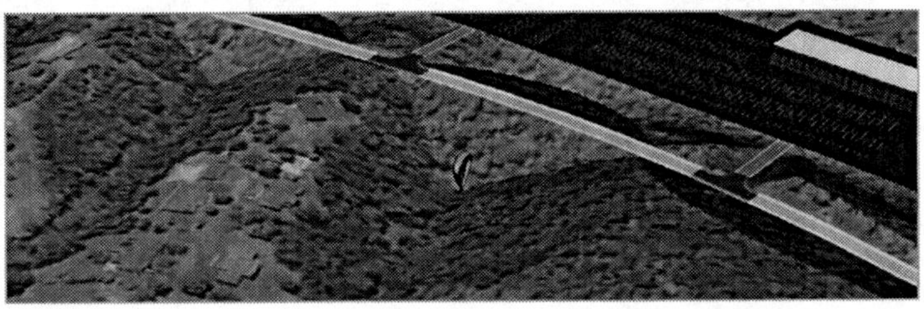

FIGURE 2.7 **A point of interest is inserted to mark the outlet point for a watershed.**

10. On the Create Watersheds asset card, set Grid Spacing to 10m and Stream Threshold to 50. Press Enter.

    You will see a series of notifications as the process takes place:

    - Preparing data
    - Uploading data to cloud
    - Running watershed analysis on cloud
    - Downloading watershed analysis results from cloud
    - Generating watersheds and stream networks

    This process should take only about a minute but will vary based on your computer speed and Internet connection. When the process

If you do not have access to cloud credits or if there is some issue with the setup of your InfraWorks 360 account, you may hit a roadblock here. If so, just skip this step and the next one and restore the Ex_2_1_End proposal to see the results.

is complete, you will see the watershed, stream network, and point of interest appear in the model. In addition, a dialog will inform you that the analysis request completed successfully along with information about the number of cloud credits utilized for the analysis (see Figure 2.8).

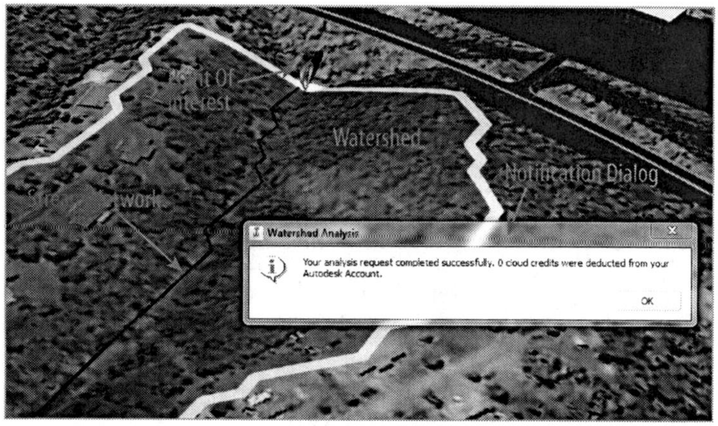

FIGURE 2.8 **A watershed analysis has just been completed.**

11. Click OK to dismiss the notification dialog. Zoom out and observe the extent of the watershed that has been created.

    The watershed clearly delineates the area that drains to this outlet point (see Figure 2.9). Later you'll use this watershed to design a culvert whose entrance will be placed at the outlet point.

If you open the Data Sources panel, you will see new entries under Streams and Watersheds.

FIGURE 2.9 **The full extent of the watershed**

You can view the results of successfully completing this exercise by selecting the Ex_2_1_End proposal.

Watershed Styles

Watersheds have their own styles, and you'll see a separate tab for them on the Style palette. They are configured exactly like coverage styles, which is by selecting a material for Fill Style and a material for Outline Style and by specifying a value for Outline Width.

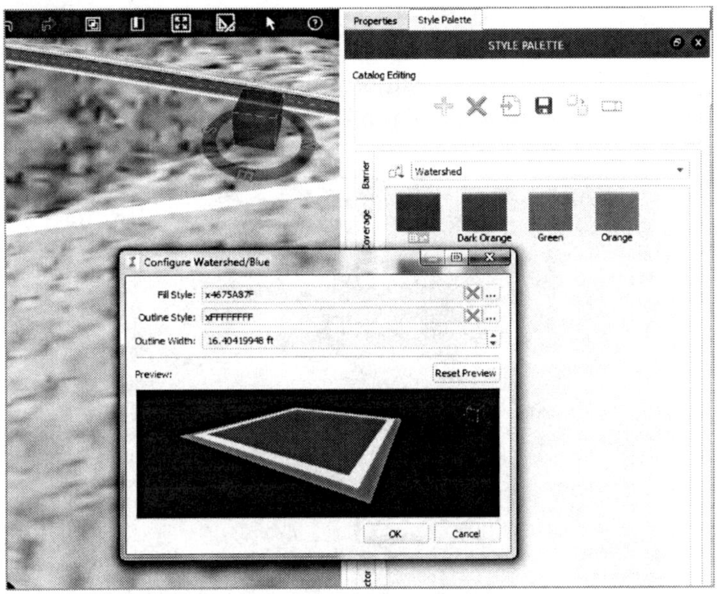

Exercise 2.2: Create Road Watersheds

In this exercise you will choose a region of the Bimsville Bypass and allow InfraWorks to automatically locate outlet points along the road as well as delineate watersheds and stream networks.

If you are continuing from the previous exercise, you can skip to step 4. Otherwise, if you haven't already done so, go to the book's web page at `www.sybex.com/go/drainageessentials` and download the files for Chapter 2. Unzip the files to the correct location on your hard drive according to the instructions in the introduction.

1. Launch InfraWorks 360.
2. On the Start Page, click Open.
3. Browse to `C:\InfraWorks Drainage Essentials\Chapter 02\` and select `Ch02 Bimsville Drainage.sqlite`. Click Open.
4. Select Ex_2_2 from the Proposals drop-down menu.
5. Click the Bookmarks icon on the Utility Bar; then click Road Watersheds to restore that bookmark.

   You are looking at a location where a portion of the Bimsville Bypass is in a fill condition and appears to be obstructing several valleys, which may carry the flow of water (see Figure 2.10).

FIGURE 2.10 **A design road is obstructing the flow of water.**

6. If the Drainage toolbar is not visible, click the Drainage icon on the main toolbar to expand the Drainage toolbar.
7. If the Design toolbar is not visible, click the Design icon on the Drainage toolbar.
8. Click Create Watersheds on the Design toolbar. If a dialog opens informing you that an Internet connection and an InfraWorks 360 account are required, click OK.

The Create Watersheds asset card will open, and a prompt will appear at the bottom of the screen instructing you to click an outlet point or select a design road.

9. Click a point on the Bimsville Bypass, the four-lane divided highway running left to right across the top of your view.

   You will now be prompted to select a region. As you move your cursor, a yellow cylinder gizmo will follow it, and a tooltip will report the station.

10. Click a point on the road such that the station value reported on the tooltip is roughly 28+00 (0+850).

    You will then be prompted for the second station, and a blue cylinder gizmo will follow your cursor while the yellow one will remain at the location of the first click.

11. Click a point near station 39+00 (1+200), as shown in Figure 2.11.

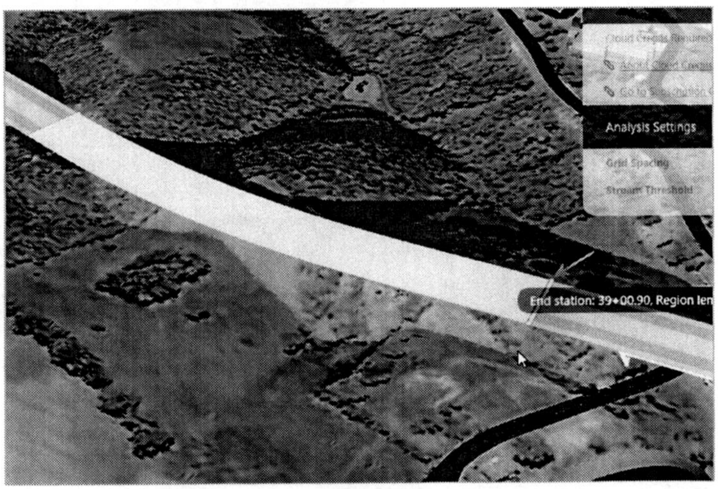

FIGURE 2.11 Choosing the region of a road for a watershed analysis (the red arrows highlight the start and end of the region)

If you do not have access to cloud credits or if there is some issue with the setup of your InfraWorks 360 account, you may hit a roadblock here. If so, just skip this step and the next one and restore the Ex_2_2_End proposal to see the results.

12. On the Create Watersheds asset card, set Grid Spacing to 10m and Stream Threshold to 50. Press Enter.

    You will see a series of notifications as the process takes place:

    - Preparing data
    - Uploading data to cloud

- Running watershed analysis on cloud
- Downloading watershed analysis results from cloud
- Generating watersheds and stream networks

This process should take only about a minute but will vary based on your computer speed and Internet connection. When the process is complete, you will see one watershed, one stream network, and a point of interest appear in the model. In addition, a dialog will inform you that the analysis request completed successfully along with information about the number of cloud credits utilized for the analysis (see Figure 2.12).

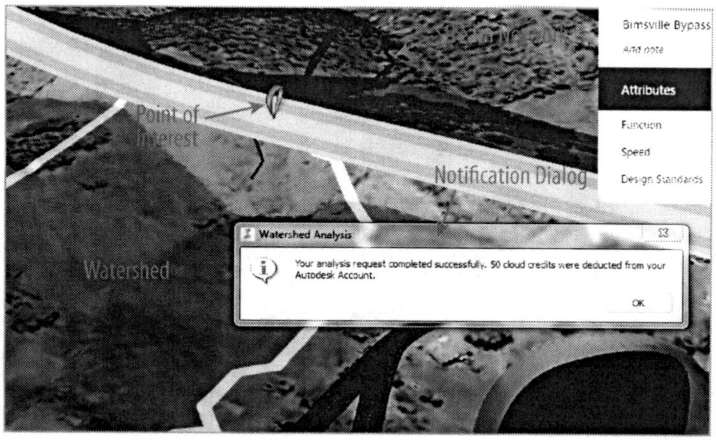

FIGURE 2.12 A watershed analysis has just been completed for a road

13. Click OK to dismiss the notification dialog. Zoom out and observe the extent of the watershed that has been created.

    Although there were at least two potential outlet points within the range you selected, only one generated a watershed and a stream network (see Figure 2.13). Other potential areas did not possess the required 5,000 square meters of upstream contributing area to generate a stream network and therefore a watershed.

    You can view the results of successfully completing this exercise by selecting the Ex_2_2_End proposal.

In the Data Sources panel, you will see new entries beneath Streams and Watersheds and a new category for Points Of Interest with its own new entry.

FIGURE 2.13 A single watershed is generated by the analysis.

Analyzing Watersheds

Once you have used Drainage Design for InfraWorks 360 to create a watershed in your model, it is fairly easy to calculate one or more peak flows for that watershed. To accomplish this, you use the Watershed asset card (see Figure 2.14) to configure the analysis and report the results.

FIGURE 2.14 The Watershed asset card

The Summary section of the Watershed asset card contains general information such as the watershed name, notes, and area. You can edit the name and notes, but the area is derived from the watershed object and cannot be edited.

The Attributes section is where the analysis is configured and the results are reported in the form of peak flows. The first few values—Channel Length, Channel Slope, and Elevations (High/Low)—are calculated automatically by the software and cannot be modified. These are key values in determining the peak flow when you choose the Rational or Regression analysis method. The next few options will vary based on which method you select.

Hydrology Method: User Defined Select User Defined when you want to use flow values that you've calculated outside InfraWorks or for times when you want to apply some "educated guesses" based on what you already know about the region and the site (see Figure 2.15).

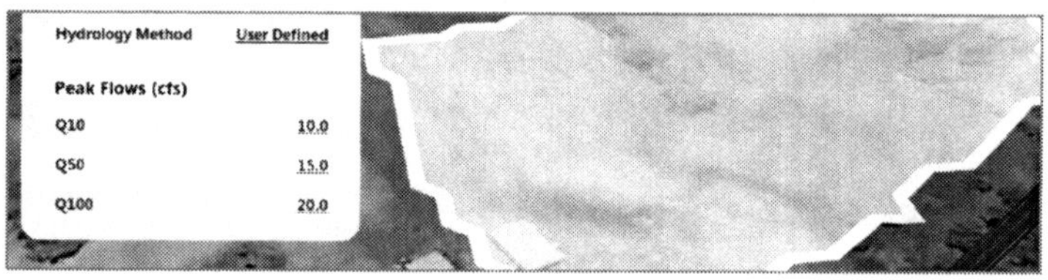

FIGURE 2.15 A sample User Defined analysis

When you choose this method, you can enter your predetermined values for Q10, Q50, and Q100. The Q100 flow will be used to size culverts.

Hydrology Method: Rational The Rational method is a commonly used method based on the characteristics of the land and historical rainfall data (see Figure 2.16).

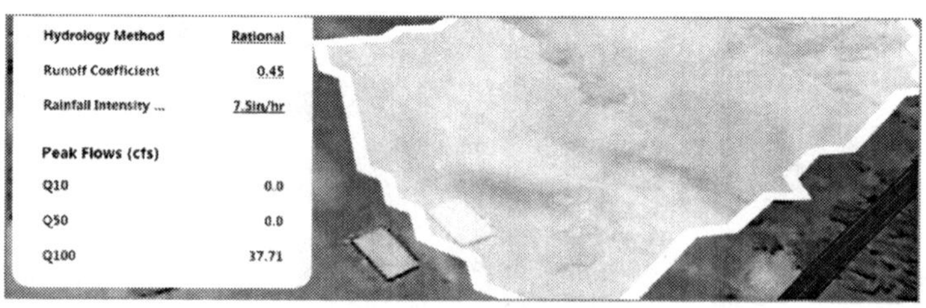

FIGURE 2.16 A sample Rational analysis

The following options are available when you choose this method:

Runoff Coefficient The runoff coefficient is a value between 0 and 1, which is lower for pervious land cover (forest) and higher for impervious (pavement and rooftops).

Rainfall Intensity The rainfall intensity is most commonly obtained from an intensity-duration-frequency (IDF) curve that has been developed for the area within which the project resides. It represents the amount of rainfall that is expected based on historical values.

Peak Flows Only the Q100 value is provided when the Rational method is chosen. It is calculated automatically and cannot be edited.

Hydrology Method: Regression The Regression method calculates flow using a regression equation. Studies have been done to derive regression equations that predict peak flows for specific regions. When you use this method, which applies only to the United States, you must choose the state and region your project is located within (see Figure 2.17).

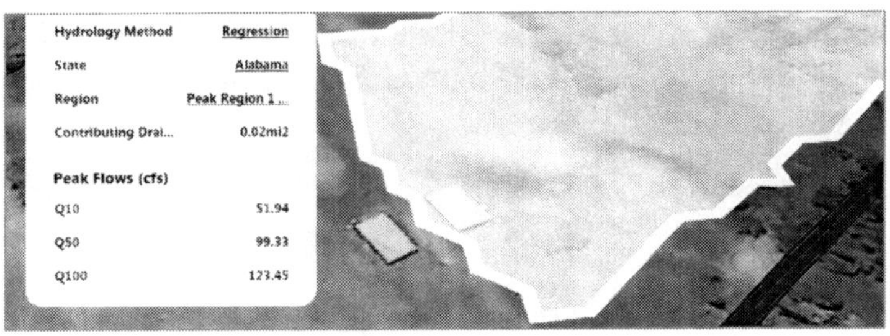

FIGURE 2.17 **A sample Regression analysis**

The following options are available when you choose this method:

State Choose the state that your project resides within.

Region Choose a region within the state. The number and names of the available regions will vary based on the state you have chosen.

Contributing Drainage Area This value is derived automatically based on the watershed itself. It cannot be edited.

Peak Flows Q10, Q50, and Q100 values are all provided. These values are calculated by the software and cannot be edited.

Exercise 2.3: Analyze Watersheds

In this exercise you will provide user-defined flows for the industrial park watershed and use both the Rational and Regression methods to analyze the Bimsville Bypass watershed.

If you are continuing from the previous exercise, you can skip to step 4. Otherwise, if you haven't already done so, go to the book's web page at `www.sybex.com/go/drainageessentials` and download the files for Chapter 2. Unzip the files to the correct location on your hard drive according to the instructions in the introduction.

1. Launch InfraWorks 360.
2. On the Start Page, click Open.
3. Browse to `C:\InfraWorks Drainage Essentials\Chapter 02\` and select `Ch02 Bimsville Drainage.sqlite`. Click Open.
4. Select Ex_2_3 from the Proposals drop-down menu.
5. Click the Bookmarks icon on the Utility Bar; then click Point Watershed to restore that bookmark.

   You are viewing a watershed that drains to the future location of a culvert. Let's say that since this design is taking place adjacent to a local road, the design standards are bit less defined, and you can simply plug in some estimated flows.
6. Click the watershed to reveal the Watershed asset card.
7. In the Attributes section, for Hydrology Method, select User Defined.

   If you are unable to edit this value, right-click the watershed and select Edit to enter edit mode.
8. For the Q10, Q50, and Q100 values under Peak Flows, enter **10** (**0.28**), **15** (**0.42**), and **20** (**0.57**), respectively, as shown previously in Figure 2.15.

   These values will be stored in the software and remain associated with this watershed. Later you will create a culvert at this location, and it will associate itself with the Q100 value.
9. Restore the Road Watersheds bookmark. Click the watershed in this area to open the Watershed asset card.
10. On the Watershed asset card, do the following:
    - For Hydrology Method, choose Regression.
    - For State, choose West Virginia.
    - For Region, choose Peak Flow Eastern Panhandle Region 2010 5033.

Note the flows that are calculated (see Figure 2.18).

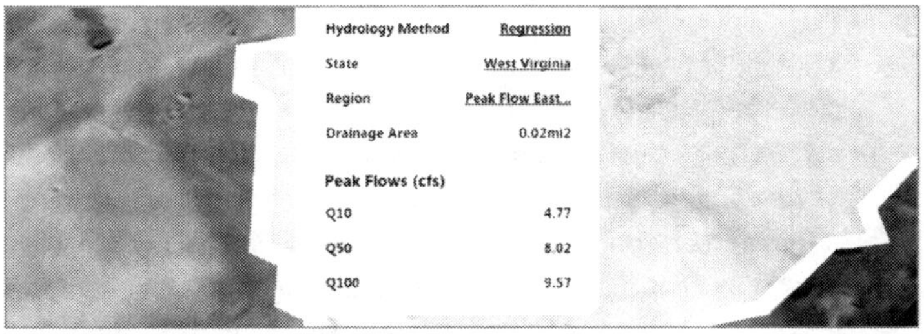

FIGURE 2.18 Results of a Regression analysis

11. On the Watershed asset card, do the following:
    - ▶ For Hydrology Method, choose Rational.
    - ▶ For Runoff Coefficient, enter **0.27**.
    - ▶ For Rainfall Intensity, enter **8 (203)**.

    Note the flows that are calculated (see Figure 2.19). Based on these results, the Rational method results yield higher flows than the Regression method results. Perhaps for additional safety, when you design the culvert later, you should consider using the more conservative Rational method results.

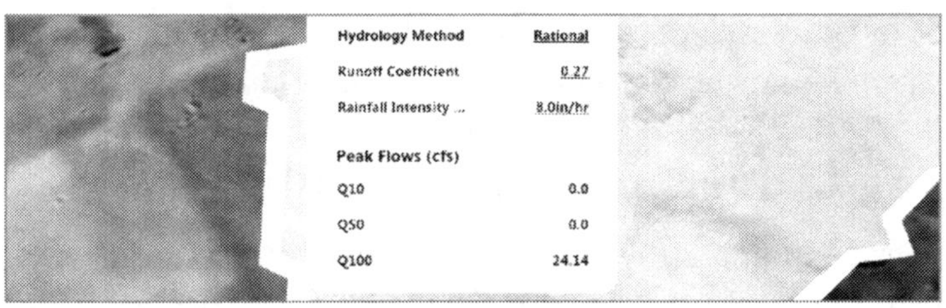

FIGURE 2.19 The results of a Rational analysis

You can view the results of successfully completing this exercise by selecting the Ex_2_3_End proposal.

How Good Are the Numbers?

Remember that InfraWorks is a tool for preliminary conceptual design. The methods applied to delineate watersheds and calculate flows produce an approximation, not a final determination. When you enter the detailed design stages of your project, you will perform these calculations in much more detail with different tools such as the Storm and Sanitary Analysis or Hydraflow tools that come with Civil 3D. However, the information that InfraWorks does provide is useful for getting key information quickly and making big decisions early in a project. For example, a quick watershed analysis may determine whether a bridge is needed vs. a culvert, which is a big cost difference that may determine feasibility, the path of a road, or the allocation of funding for other purposes. As you have seen, Drainage Design for InfraWorks 360 can provide these key numbers in minutes, which is perfect for a preliminary design or proof of concept. A detailed hydrologic and hydraulic study can take months to prepare, and that is appropriate for the detailed design stage of the project.

Creating Culverts

You can create culverts using Drainage Design for InfraWorks 360 in two ways. One way is to let InfraWorks create the culverts automatically, and the other is to create them manually. You will study both methods in this section.

Creating Culverts Automatically

InfraWorks can automatically create culverts wherever a road crosses a water area (a linear water area, not a closed shape like a pond or lake) or a stream. At first, this may not seem all that impressive until you consider all that is done automatically and the relationships that are established as a result. For example, not only does InfraWorks detect where a water area or stream crosses a road, it also calculates the location where the embankments of the road intersect the stream or water area to determine the exact 3D placement of the culvert inlet and outlet. If a watershed is present at the culvert inlet, InfraWorks will automatically use the Q100 peak flow and size the culvert based on its shape, slope, and other factors. As if that isn't enough, InfraWorks will maintain the relationship between the culvert, location of the road embankments, location of the stream or water area, and peak flow of the watershed. If anything changes, the culvert will be updated automatically. You will learn about manual culvert placement in the next section, but based on everything that was just mentioned, you should seek to have InfraWorks automatically design your culverts whenever possible.

To automatically create culverts, you simply select a design road and right-click it while in edit mode. When you do this, the shortcut menu will contain an Add Culverts command (see Figure 2.20). When you click this command, InfraWorks will insert culverts wherever a stream or water area intersects the road.

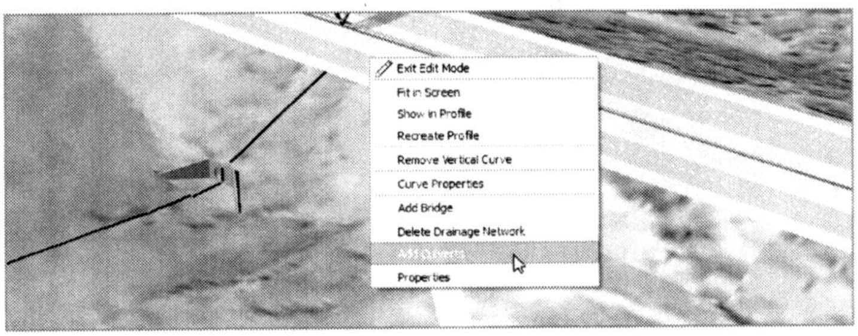

FIGURE 2.20 The Add Culverts command

Exercise 2.4: Create a Culvert Automatically

In this exercise you will have InfraWorks automatically create a culvert beneath the Bimsville Bypass at the location of a previous calculated watershed. You will then make an edit to the road to observe the dynamic behavior of the culvert.

If you are continuing from the previous exercise, you can skip to step 4. Otherwise, if you haven't already done so, go to the book's web page at `www.sybex.com/go/drainageessentials` and download the files for Chapter 2. Unzip the files to the correct location on your hard drive according to the instructions in the introduction.

1. Launch InfraWorks 360.
2. On the Start Page, click Open.
3. Browse to `C:\InfraWorks Drainage Essentials\Chapter 02\` and select `Ch02 Bimsville Drainage.sqlite`. Click Open.
4. Select Ex_2_4 from the Proposals drop-down menu.
5. Click the Bookmarks icon on the Utility Bar; then click Road Watersheds to restore that bookmark.

In this version of the model, the stream network was removed, and a more well-defined stream feature was drawn using the Rivers command. This new water feature flows through the valley to a small pond to the southwest of where it crosses beneath the road (see Figure 2.21).

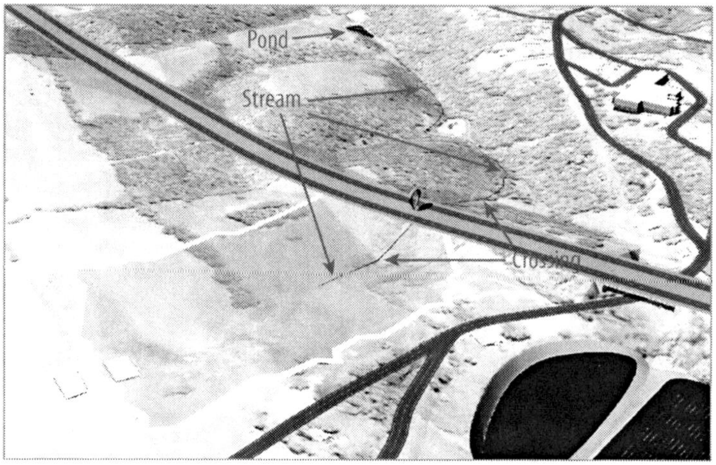

FIGURE 2.21 An existing stream crossing beneath the Bimsville Bypass

6. Click the road to select it; then right-click and select Add Culverts.

   A new culvert will be created that begins where the toe of the eastern embankment meets the stream and ends where the toe of the western embankment meets the stream.

If you do not see the Add Culverts command, you may not be in edit mode. If that is the case, right-click and select Edit before performing this step.

7. Restore the Culvert Upstream bookmark to view the entrance to the culvert (see Figure 2.22).

FIGURE 2.22 A view of the upstream end of the culvert

8. Restore the bookmark named Culvert Underground to view the culvert from beneath the road (see Figure 2.23).

FIGURE 2.23 An underground view of the culvert

Underground View Settings

Viewing from underground works best if you disable the Lock Mouse Navigation Above Ground option and enable the Show Backfaces option. You can find both of these settings in the Application Options dialog in the Navigation and 3D Graphics sections, respectively.

9. Restore the bookmark named Culvert Downstream.

   You are now looking at the downstream end of the culvert. Take note of the location of the culvert exit. You are about to widen the Bimsville Bypass and watch the dynamic behavior as the culvert automatically changes its own design.

10. Click the Bimsville Bypass to open the Road asset card. On the Road asset card, change Edit Mode to Lanes Backward.

11. Change the value for Lanes to 3 and watch as the road widens, the embankment pushes to the west, and the end of the culvert moves westward to match.

    In Figure 2.24 the image on the left reflects two lanes (before), and the image on the right reflects three lanes (after).

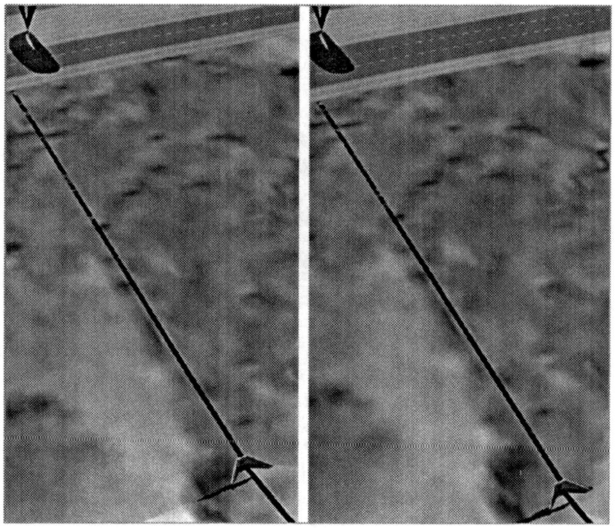

FIGURE 2.24 The culvert has updated its design by shifting the downstream end to match the new embankment location.

You can view the results of successfully completing this exercise by selecting the Ex_2_4_End proposal.

Creating Culverts Manually

As you have learned, InfraWorks will automatically place a culvert wherever a water area or stream crosses a design road. At times you may need to create culverts where there is no road, stream, or water area. In this case, you can use the Culverts tool and hand pick where a culvert starts and ends.

The Culverts tool is on the Design toolbar of the Drainage Design tools, as shown in Figure 2.25. After launching the command, you simply click two points representing either end of the culvert. InfraWorks will automatically place the invert of each end of the culvert to match the ground elevation of the model. The end of the culvert with the higher elevation will automatically be designated as the upstream end, and the lower elevation will be the downstream end, regardless of which direction you draw the culvert. After picking the second point, the Culvert asset card will open automatically. If the upstream end of the culvert is near the outlet point of a watershed, the calculated Q100 flow will automatically be applied to the culvert, and it will be sized accordingly.

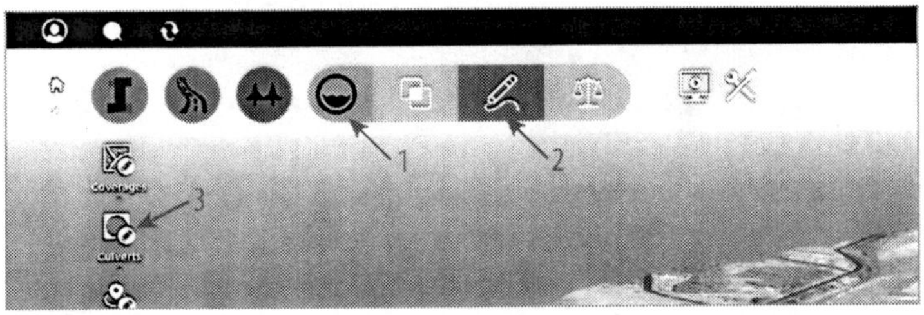

FIGURE 2.25 To launch the Culverts tool, you can click the icons in the order indicated by the numbers.

Culverts in Profile

If you view a design road in the Profile View panel and a culvert is present for the road, it will show up in the profile in the form of a blue diamond symbol. If you click the blue diamond, a label will appear next to it that reports the station and elevation where the culvert crosses the road centerline. If you right-click the diamond symbol, you can click the Show In Canvas command to zoom to the culvert in the model.

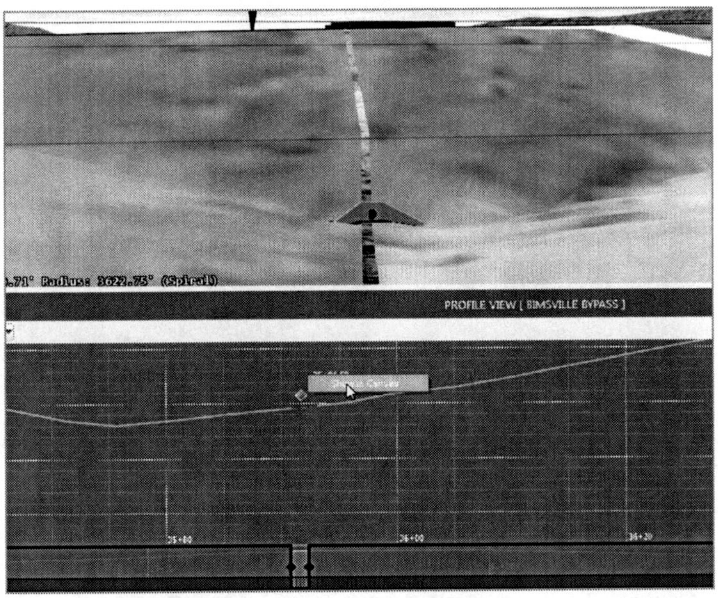

Exercise 2.5: Create Culverts Manually

In this exercise you will create two new culverts that allow runoff to flow beneath the industrial park development and continue westward along its natural course.

If you are continuing from the previous exercise, you can skip to step 4. Otherwise, if you haven't already done so, go to the book's web page at `www.sybex.com/go/drainageessentials` and download the files for Chapter 2. Unzip the files to the correct location on your hard drive according to the instructions in the introduction.

1. Launch InfraWorks 360.
2. On the Start Page, click Open.
3. Browse to `C:\InfraWorks Drainage Essentials\Chapter 02\` and select `Ch02 Bimsville Drainage.sqlite`. Click Open.
4. Select Ex_2_5 from the Proposals drop-down menu.
5. Click the Bookmarks icon on the Utility Bar; then click Culvert 2 to restore that bookmark.

   You are viewing the potential location of a culvert with the upstream end at the blue ribbon marker in the top of your view and the downstream end at the blue flag marker at the bottom of your view (see Figure 2.26).

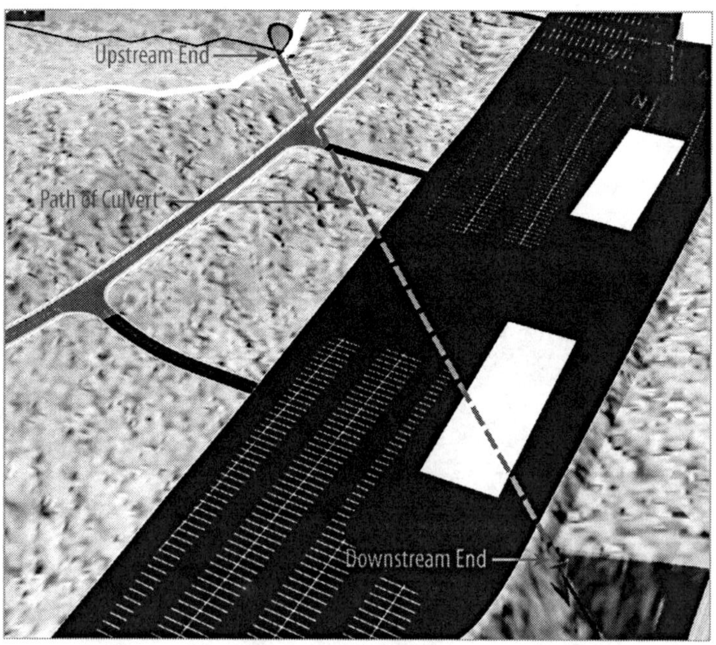

FIGURE 2.26 The potential location for a culvert

6. If the Drainage toolbar is not expanded, click the Drainage icon to expand it.
7. If the Design toolbar is not visible, click the Design icon on the Drainage toolbar.
8. Click Culverts on the Design toolbar.
9. Zoom in to the blue ribbon marker at the top of your view and click a point near its base.
10. Zoom out until you can see the blue flag marker to the northwest (down and to the right if you haven't changed your view perspective since selecting the bookmark); then zoom into this marker and pick a point near its base.

    After a pause, a new culvert will be created.
11. Delete the flag marker at the downstream end of the culvert. The downstream end of the culvert should appear similar to Figure 2.27.

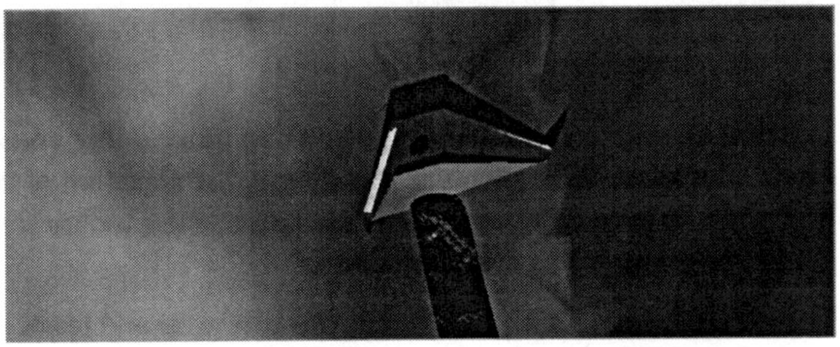

FIGURE 2.27 The downstream end of a newly created culvert

If you do not see two orange flags, your graphics settings may be too low to display both of them. Try zooming in or adjusting your settings.

12. Restore the bookmark named Culvert 3.

    You are looking at the northern end of a parking lot with an orange flag on either side of the raised area that the parking lot rests on. In the next few steps you will create a culvert from one flag to the other.
13. Launch the Culverts tool and pick near the base of each orange flag. Delete the flags when you are finished.

    Another new culvert has been created.

    You can view the results of successfully completing this exercise by selecting the Ex_2_5_End proposal.

Editing Culverts

As with all design, the first iteration is never the last, so a good set of editing tools is always critical. When it comes to culverts, InfraWorks has some tools that allow you to edit culverts graphically and through the use of the Culvert asset card. I'll discuss both of those approaches in this section.

Editing Culverts Graphically

As you have seen with other InfraWorks objects, culverts have a set of gizmos that enable specific types of graphical editing (see Figure 2.28).

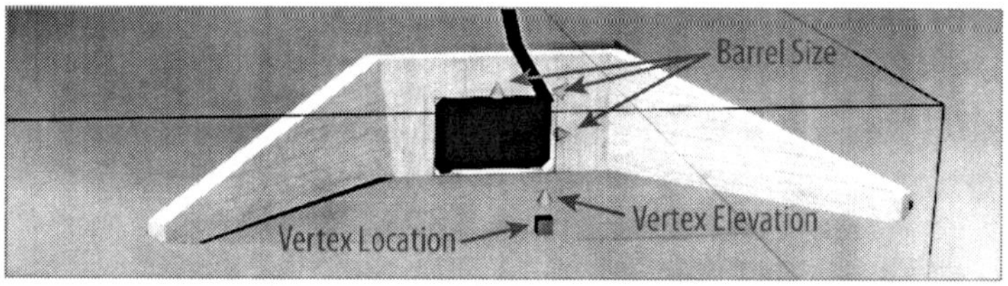

FIGURE 2.28 Culvert gizmos

Vertex Location Gizmo With this gizmo you can move either end of the culvert to a new location. When you use this gizmo, the elevation of the culvert adjusts automatically to match the terrain. You can use the tooltip to enter specific coordinates, although this is seldom done.

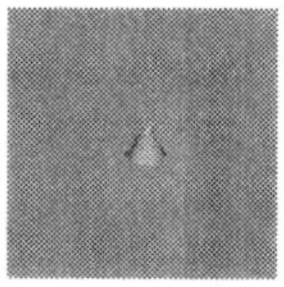

Vertex Elevation Gizmo With this gizmo you can raise and lower the elevation of either end of the culvert. If you use the vertex location gizmo to edit the location of the culvert after adjusting the elevation with this gizmo, the elevation will reset itself to match the terrain. You can use the tooltip to enter a specific elevation, but be aware that this is not the invert elevation of the barrel. It will be close but not exactly the same.

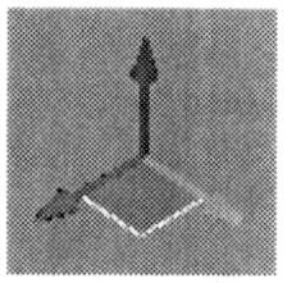

The Move Gizmo The move gizmo has four parts: three axis arrows (x, y, and z) and a square in the x-y plane. If you click and drag an arrow, you will move the entire culvert along that axis. If you click and drag the square, you will move the culvert in the x-y plane only. You can use tooltips to enter the change in a given axis, which can be especially handy in the z-axis if, for instance, you'd like to raise the culvert by a certain value.

The Barrel Size Gizmo You can click and drag this gizmo to change the barrel size of a culvert. For circular culverts there will be only one gizmo for the diameter. For box culverts there will be three gizmos. The one on top will change the height, the one on the side will change the width, and the one at the corner will change the height and width simultaneously (see Figure 2.28). When you move any of the barrel size gizmos, they will "snap" to specific sizes to match what is commonly used in the industry.

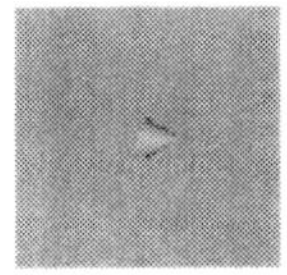

Exercise 2.6: Edit Culverts Graphically

In this exercise you will use gizmos to edit two of the culverts in the model.

If you are continuing from the previous exercise, you can skip to step 4. Otherwise, if you haven't already done so, go to the book's web page at `www.sybex.com/go/drainageessentials` and download the files for Chapter 2. Unzip the files to the correct location on your hard drive according to the instructions in the introduction.

1. Launch InfraWorks 360.
2. On the Start Page, click Open.
3. Browse to `C:\InfraWorks Drainage Essentials\Chapter 02\` and select `Ch02 Bimsville Drainage.sqlite`. Click Open.
4. Select Ex_2_6 from the Proposals drop-down menu.
5. Click the Bookmarks icon on the Utility Bar; then click Culvert 2 Upstream to restore that bookmark.

   You are viewing the upstream end of a culvert. The culvert opening doesn't quite match up with the end of the stream.
6. Click the culvert to display its gizmos. If the gizmos do not appear, right-click and select Edit.
7. Click the purple box gizmo and drag it to the base of the blue flag marker.

   After a pause, the culvert entrance will line up with the end of the stream. Notice that the vertex location gizmos for a culvert are not located at the centerline of the barrel opening. Instead, they are located where the centerline of the barrel(s) intersects the front edge of whatever end treatment is in use. In this case, it is the front edge of the floor of the wingwall structure (see Figure 2.29).

FIGURE 2.29 The culvert entrance has been repositioned using the vertex location gizmo.

8. Restore the bookmark named Culvert 2 Downstream.
9. If the culvert gizmos are not visible, click the culvert.
10. Click the vertex elevation gizmo and enter **695 (212)** in the tooltip.

    After a pause, the culvert will update slightly. Using the tooltip is a handy way to provide an even value for the elevation of the culvert exit. However, please note that this elevation is applied at the gizmo, not at the exact end of the barrel.

11. Restore the bookmark named Culvert 3 Upstream. Click the culvert at this location to reveal its gizmos.
12. Click the barrel size gizmo and drag it to the right until the tooltip reads Size: 36in (900mm), as shown in Figure 2.30.

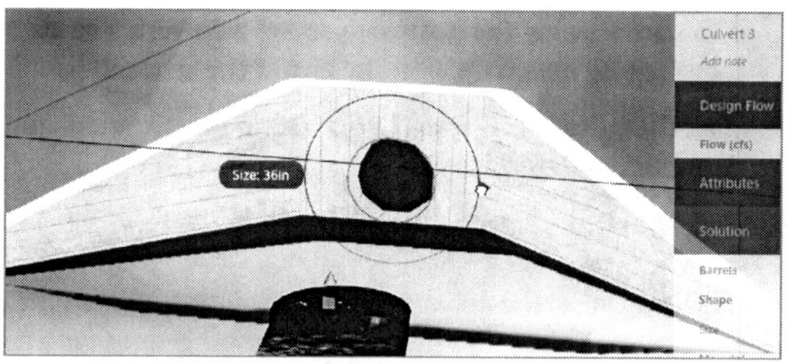

FIGURE 2.30 Resizing a culvert barrel

After a pause, the barrel size and structure will update. You can view the results of successfully completing this exercise by selecting the Ex_2_6_End proposal.

Using the Culvert Asset Card

The design of a culvert can also be edited using the Culvert asset card (see Figure 2.31). You open this asset card by simply clicking a culvert in the model. To change the values on the Culvert asset card, edit mode must be activated. Following this paragraph is a listing of the sections and options contained on the Culvert asset card. For now, I will not discuss the Analysis section of the asset card and reserve that for a bit later.

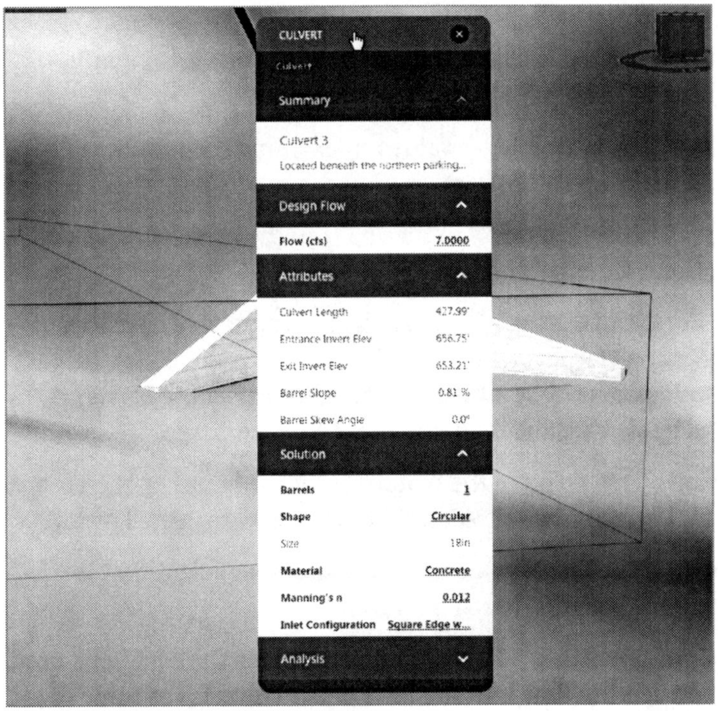

FIGURE 2.31 The Culvert asset card

Summary The Summary section contains general information: the name of the culvert and a space to provide a note.

Design Flow The Design Flow section will contain one of two possible options: Flow or Watershed Flow. If the name of the option is Flow, this indicates there is no upstream watershed associated with the culvert and the value can be edited directly. If it is named Watershed Flow, the value is derived automatically and cannot be edited directly.

Attributes The Attributes section contains a list of geometric properties that cannot be edited.

- Culvert Length
- Entrance Invert Elevation
- Exit Invert Elevation
- Barrel Slope
- Barrel Skew Angle

Solution The Solution section is where the design of the culvert is configured. The options are as follows:

Barrels This is an integer value representing the number of barrels that make up the culvert.

Shape This option refers to the shape of the barrel(s). There are two options: Circular and Box.

Size For circular barrels, the size will be represented by one value (diameter), and for rectangular it will display as two (width × height). These values cannot be edited directly. The size of the barrels is changed by clicking and dragging the barrel size gizmo.

Material This refers to the material that the barrel or barrels are made of. There are two choices: Concrete and Corrugated Metal.

Manning's N The N value is a roughness coefficient that is used for sizing the barrel(s) based on flow.

Inlet Configuration This option determines the equations used when considering flow into the inlet of the culvert. For some of the choices, it also controls the configuration of the headwall and wingwalls. This is based on methodology developed by the U.S. Federal Highway Administration.

The choices for a circular culvert are as follows:

- Square Edge with Headwall
- Groove End in Headwall
- Groove End Projecting

The choices for a box culvert are as follows:

- 30deg to 75deg Wingwall Flares
- 90deg and 15deg Wingwall Flares
- 0deg Wingwall Flares
- 45deg Wingwall Flare d = 0.043D
- 18deg to 33.7deg Wingwall Flare d = 0.083D
- 90deg Headwall With 3/4″ Chamfers
- 90deg Headwall with 45deg Bevels
- 90deg Headwall with 33.7deg Bevels
- 3/4″ Chamfers; 45deg Skewed Headwall
- 3/4″ Chamfers; 30deg Skewed Headwall
- 3/4″ Chamfers; 15deg Skewed Headwall
- 45deg Bevels; 10deg-45deg Skewed Headwall

Exercise 2.7: Modify a Culvert Using the Culvert Asset Card

In this exercise you will change the configuration of a culvert using the Culvert asset card.

If you are continuing from the previous exercise, you can skip to step 4. Otherwise, if you haven't already done so, go to the book's web page at `www.sybex.com/go/drainageessentials` and download the files for Chapter 2. Unzip the files to the correct location on your hard drive according to the instructions in the introduction.

1. Launch InfraWorks 360.
2. On the Start Page, click Open.
3. Browse to `C:\InfraWorks Drainage Essentials\Chapter 02\` and select `Ch02 Bimsville Drainage.sqlite`. Click Open.
4. Select Ex_2_7 from the Proposals drop-down menu.
5. Click the Bookmarks icon on the Utility Bar; then click Culvert 2 Downstream to restore that bookmark.

   You are viewing the downstream end of a culvert.

Where's My Watershed?

You may have noticed that it appears that the watershed at the upstream end of this culvert has been removed. Actually a style named Invisible has been applied to it. In this style, the Fill Style setting uses a color with an Alpha Channel value of 0 (transparent) and Outline Width is set to 0. If you pass your cursor over the watershed, you will see it highlight. This is a handy way to keep your watersheds in the model but hide them for visualization purposes.

6. Click the culvert to display the Culvert asset card.
7. On the Culvert asset card, click Add Note and enter **Culvert beneath parking lot number 4**.
8. Change the value for Barrels to 3.
9. Change the value for Shape to Box.
10. Change the value for Inlet Configuration to 90deg and 15deg Wingwall Flares.
11. Press Esc to clear the selection of the culvert. Zoom in and study the new configuration of the culvert.

    The culvert should look similar to Figure 2.32.

FIGURE 2.32 A newly configured culvert

You can view the results of successfully completing this exercise by selecting the Ex_2_7_End proposal.

Analyzing Culverts

In addition to placing, sizing, and modifying culverts, you can also perform an analysis of the hydraulics of a culvert. This is done using the Analysis section of the Culvert asset card. This section has the following options:

Show Analysis Results This option will toggle the graphical analysis that hovers above the culvert when you select it (see Figure 2.33). This analysis will report upstream and downstream conditions as well as flow per barrel. In addition, it will provide an interactive graphic of flow depth along the length of the culvert. As you move your cursor along the blue 3D graphic, tooltips will report the distance along the culvert vs. the depth of flow.

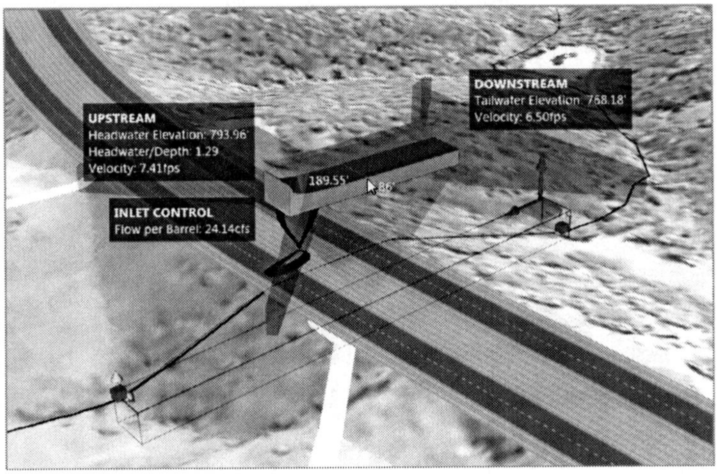

FIGURE 2.33 Graphical culvert analysis

Tailwater Condition This option controls the approach used to calculate the tailwater elevation. The choices are as follows:

- (dc+D)/2
- Crown
- Critical
- Normal
- User Defined

The User Defined option allows you to directly enter a value for Tailwater Elevation. The others provide calculated values that you cannot change directly. In the first option, dc represents critical depth, and D represents the diameter of the barrel.

Tailwater Elevation This is the calculated elevation of the water just beyond the culvert outlet. Unless Tailwater Condition is set to User Defined, this value cannot be edited but is instead calculated or taken directly from the culvert geometry. There is a built-in minimum value for this, and if you enter a value that is lower, InfraWorks will automatically update it to the minimum.

In addition to providing the analysis results graphically and on the Culvert asset card, Drainage Design for InfraWorks 360 also enables you to export your culvert analysis results to a file. This is done by right-clicking the culvert while in edit mode and selecting Export Analysis To CSV, as shown in Figure 2.34.

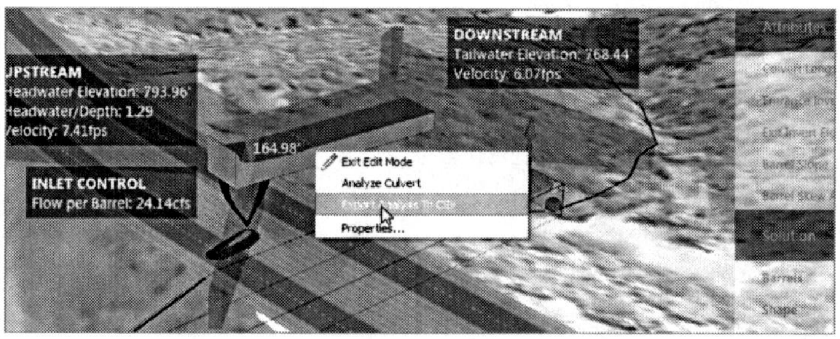

FIGURE 2.34 Exporting a culvert analysis

Exercise 2.8: Analyze a Culvert

In this exercise you will perform a culvert analysis and export the results to a CSV file.

If you are continuing from the previous exercise, you can skip to step 4. Otherwise, if you haven't already done so, go to the book's web page at www.sybex.com/go/drainageessentials and download the files for Chapter 2. Unzip the files to the correct location on your hard drive according to the instructions in the introduction.

1. Launch InfraWorks 360.
2. On the Start Page, click Open.
3. Browse to `C:\InfraWorks Drainage Essentials\Chapter 02\` and select `Ch02 Bimsville Drainage.sqlite`. Click Open.

4. Select Ex_2_8 from the Proposals drop-down menu.
5. Click the Bookmarks icon on the Utility Bar; then click Culvert Upstream to restore that bookmark.

   You are viewing the upstream end of a culvert that passes beneath the Bimsville Bypass.
6. Click the culvert to display the Culvert asset card.
7. On the Culvert asset card, if it is not already on, toggle on Show Analysis Results.
8. Restore the bookmark named Culvert Analysis.

   You should be viewing a graphical analysis of the culvert, as shown previously in Figure 2.33.
9. Change the value for Tailwater Condition to Crown.

   The value for Tailwater Elevation updates to match the crown of the barrel outlet, which is the invert plus the diameter or 766.19 + 27/12 (233.54 + 675/1000).
10. Right-click the culvert and select Export Analysis To CSV.
11. Browse to the Chapter 02 class data folder and enter a filename of **Culvert 1 Analysis**. Click Save.

    After a pause, the newly created CSV file will be available to open.
12. Open the resulting CSV file and study the results. The results will look best if opened in a spreadsheet program. Figure 2.35 shows a section of the resulting data viewed in a spreadsheet program.

No of Barrels	Entrance Type	Size	Flow	Storm Event	Critical Depth	Normal Depth	DownStream Velocity	Dc
		in	cfs	Years	ft	ft	ft/s	ft
1	Square Edge with Headwall	18	0	10	0	0	0	
1	Groove End in Headwall	18	0	10	0	0	0	
1	Groove End Projecting	18	0	10	0	0	0	
2	Square Edge with Headwall	18	0	10	0	0	0	
2	Groove End in Headwall	18	0	10	0	0	0	
2	Groove End Projecting	18	0	10	0	0	0	
1	Square Edge with Headwall	18	0	50	0	0	0	
1	Square Edge with Headwall	18	24.14	100	1.48	1.12	13.66	
1	Groove End in Headwall	18	0	50	0	0	0	
1	Groove End in Headwall	18	24.14	100	1.48	1.12	13.66	
1	Groove End Projecting	18	0	50	0	0	0	
1	Groove End Projecting	18	24.14	100	1.48	1.12	13.66	
2	Square Edge with Headwall	18	0	50	0	0	0	
2	Square Edge with Headwall	18	24.14	100	1.32	0.71	6.83	
2	Groove End in Headwall	18	0	50	0	0	0	
2	Groove End in Headwall	18	24.14	100	1.32	0.71	6.83	
2	Groove End Projecting	18	0	50	0	0	0	
2	Groove End Projecting	18	24.14	100	1.32	0.71	6.83	

FIGURE 2.35 An exported culvert analysis

You can view the results of successfully completing this exercise by selecting the Ex_2_8_End proposal. You can find the CSV export file in your Chapter 02 class data folder; it is named Ex_2_8_End.csv (Ex_2_8_End_Metric.csv).

Where Do the Numbers Come From?

It is important to remember that InfraWorks and its add-on modules are for preliminary and conceptual design. Engineering principles and methodologies are applied, but the designs that InfraWorks creates are approximate and not finalized. Drainage Design for InfraWorks 360 uses a combination of assumptions, accepted methodologies, and common practices to give you a rough design and some ideas of key analysis numbers so that you can make important decisions early. For example, if you place a culvert beneath a road and the analysis provided by InfraWorks shows that the road is being overtopped, it gives you an idea that the culvert design must be modified, the road profile should be elevated, or perhaps a bridge must be used in lieu of a culvert. It does not give you a final design that you can submit for construction or to review agencies for approval. For more information on some of the methodologies applied to culverts, research the HDS-5 methodology of the United States Federal Highway Administration.

Working with Drainage Networks

A drainage network is a system of inlets, pipelines, and manholes that collects and conveys drainage to a safe location where it can be discharged. Drainage Design for InfraWorks 360 provides a set of tools for creating, modifying, and analyzing drainage networks. In this section you'll look at understanding the purpose and function of drainage networks as well as how to add them to your model and work with them to achieve your design goals.

Understanding Drainage Networks

The components of a drainage network are similar to the pipelines and pipe connectors that you find in basic InfraWorks. There are a few differences that are described as follows:

Automatic Creation You can automatically create a drainage network on a road using the Add Drainage Network command found in the design road right-click menu. There is no such command for the pipeline and pipe connector objects created by the basic InfraWorks tools.

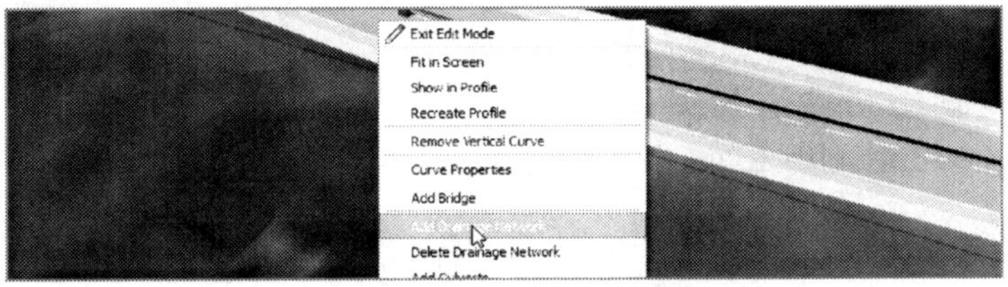

Association with a Road When you create a drainage network automatically using the Add Drainage Network command, the objects in the network remain associated with the road, and if the alignment of the road changes, the objects will update accordingly.

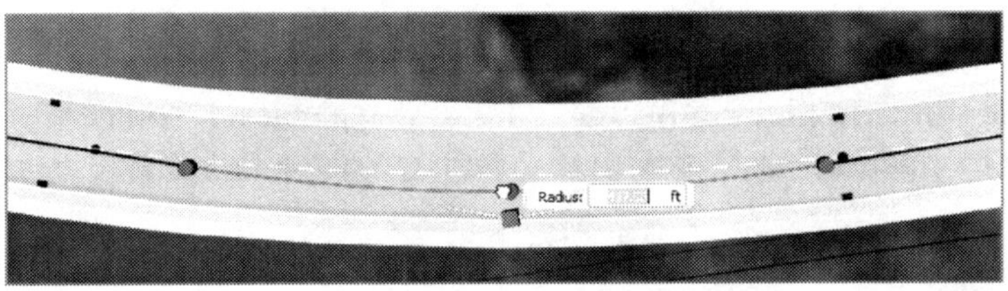

Design Rules The inlets, manholes, and pipelines you create using the drainage network tools will have depths and slopes applied automatically based on assumptions, common practices, and other design rules. For this reason you will find that much less editing is required after first creating these objects when compared to the use of basic pipelines and pipe connectors.

Asset Cards Drainage Design for InfraWorks 360 provides asset cards for inlet, manhole, and pipeline objects. These are not available for basic pipelines or pipe connectors.

Detailed Structures The inlets and manholes that are created when you use the drainage network tools are much more detailed than those created with the basic pipeline and pipe connector tools.

Limited Choices One limitation of drainage networks is that there are no options for changing the appearance or geometry of the inlets and manholes. You cannot apply the standard pipe connector styles to these objects. You *can* apply pipeline styles to pipelines created with the drainage network tools.

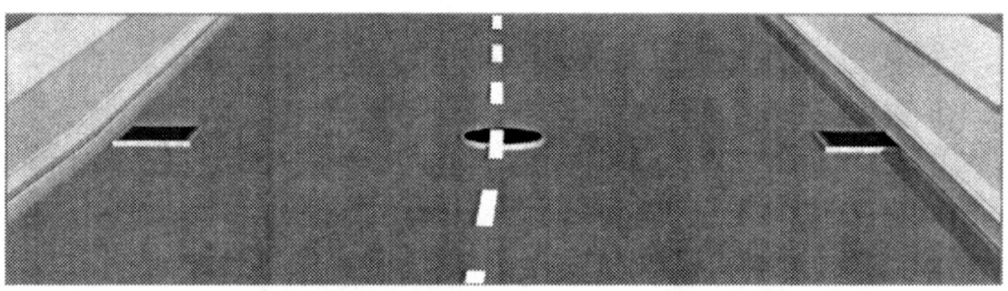

Quantities The objects in a drainage network will show up in on the Quantities asset card when you launch the Quantities command found on the Analyze toolbar of the Drainage Design intelligent tools.

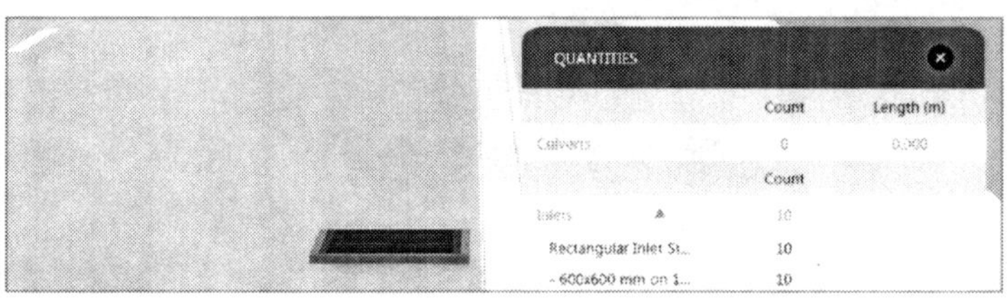

Creating Drainage Networks

You can create drainage networks with Drainage Design for InfraWorks 360 using two methods: automatically on a road or manually wherever you like. To create a drainage network automatically on a road, you first select the road while edit mode is on and then right-click and select Add Drainage Network (see Figure 2.36). InfraWorks will place inlets at low points and at intervals along slopes according to built-in design rules. This command is available only if the road style includes curbs. You should also know that there is a Delete Drainage Network command on the same right-click menu that will remove all of the inlets, manholes, and pipelines within a drainage network that is associated with the road you have selected.

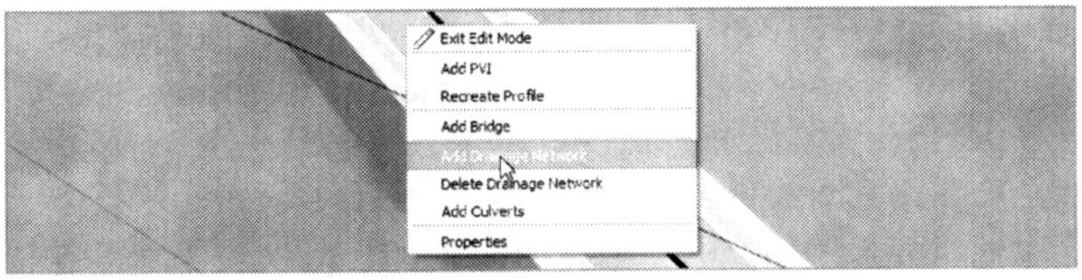

FIGURE 2.36 **The Add Drainage Network command**

To create drainage networks manually, you click the Pavement Drainage icon located on the Design toolbar of the Drainage Design intelligent tools (see Figure 2.37). After launching this tool, you can immediately begin clicking points in the model to insert inlets. As you create inlets, pipes will connect them automatically. You can double-click the last inlet location to complete the command.

FIGURE 2.37 **You can click the icons in the order shown to launch the Pavement Drainage command.**

If you want other options, you can right-click while the command is active, and a right-click menu will allow you to choose which components you would like to be inserted as you click the model (see Figure 2.38). The choices include Inlets And Pipelines, Manholes And Pipelines, Inlets Only, and Manholes Only. If you right-click after clicking one or more points, you will get the additional options of Show High And Low Points and End Run.

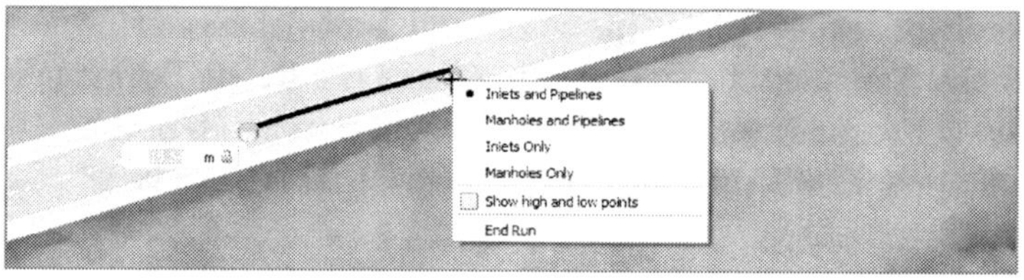

FIGURE 2.38 The right-click menu for the Pavement Drainage command

Show High And Low Points is handy when manually placing inlets because it will provide temporary markers and labels at high and low points for all design roads in the model (see Figure 2.39). Also, you can click inlets and manholes already in the model, and rather than creating a new structure, InfraWorks will connect a pipe to the structure and extend it to the new structure that you create with the next click.

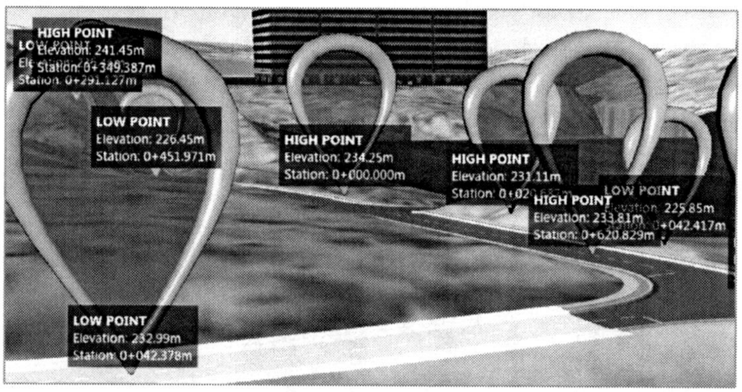

FIGURE 2.39 Markers and labels indicate high and low points in the model.

Editing Drainage Networks

The manholes, inlets, and pipelines that make up drainage networks can be edited in the same ways as basic pipelines and pipe connectors. They have the same gizmos and exhibit the same graphical editing behavior. Although there are asset cards for inlets, manholes, and pipelines, none of the values can be edited (except for the name and a note). The Properties palette for drainage network components is similar to what you will find for pipelines and pipe connectors with a few restrictions. For example, the Manual Style and Rule Style properties are not available for an inlet or manhole because their styles cannot be changed.

Exercise 2.9: Work with Drainage Networks

In this exercise, you will use the automatic and manual tools to design a drainage network for one of the industrial park roads. You will also use the Quantities command to display results for the drainage structures associated with this road as well as the main industrial park road.

If you are continuing from the previous exercise, you can skip to step 4. Otherwise, if you haven't already done so, go to the book's web page at `www.sybex.com/go/drainageessentials` and download the files for Chapter 2. Unzip the files to the correct location on your hard drive according to the instructions in the introduction.

1. Launch InfraWorks 360.
2. On the Start Page, click Open.
3. Browse to `C:\InfraWorks Drainage Essentials\Chapter 02\` and select `Ch02 Bimsville Drainage.sqlite`. Click Open.
4. Select Ex_2_9 from the Proposals drop-down menu.
5. Click the Bookmarks icon on the Utility Bar; then click Drainage Network to restore that bookmark.

   You are viewing some of the roads within the industrial park. The road you will be working on is the one heading eastward and ending at the parking lot on the left side of your view.

6. Click the road to select it; then right-click and select Add Drainage Network. If the Add Drainage Network command is not visible, make sure edit mode is on and that the Edit Mode property in the Road asset card is set to Geometry.

   A series of inlets, manholes, and pipelines has been created along the road.

7. Restore the bookmark named Low Point. If you have cleared the selection of the road, select it once again to view the gizmos.

   Notice that inlets were automatically placed at the low point as indicated by the light blue cylindrical gizmo (see Figure 2.40).

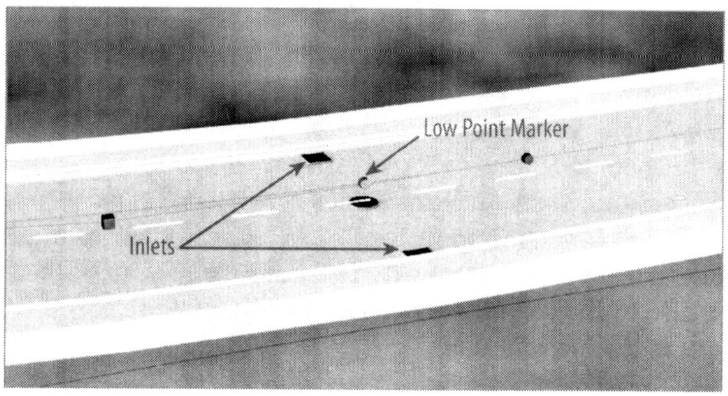

FIGURE 2.40 Inlets automatically placed at a low point

8. Restore the bookmark named Edit Drainage.

   You are viewing a location where a parking lot entrance intersects with the road. It would be ideal to intercept the drainage flowing out of this entrance before it enters the road. For this reason you will place an inlet at the end of the lower curb return and tie it into the system created earlier.

9. If the Drainage Design toolbar is not already expanded, click the Drainage Design icon.

10. If the Design toolbar is not visible, click the Design icon.

11. On the Design toolbar, click Pavement Drainage.

12. Pick a point along the curb near the end of the lower curb return and then click a point on the inlet down the road to the right. Figure 2.41 shows the two points you should click.

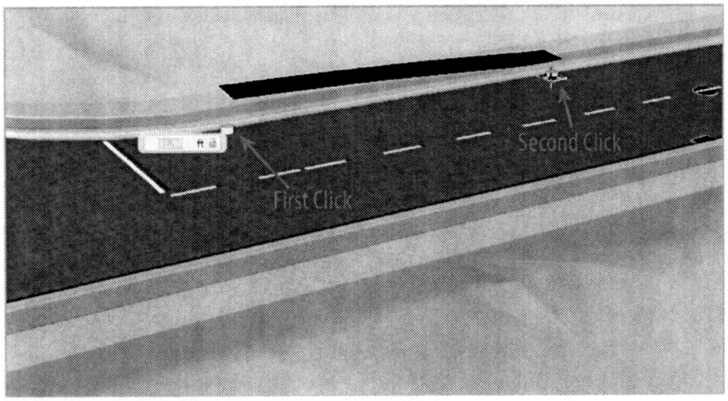

FIGURE 2.41 **Manually adding an inlet and pipeline to a drainage network**

13. Right-click and then check the box next to Show High And Low Points. Study the markers and labels you see.

    Blue ribbon markers and accompanying labels show you where the low points and high points are and what their elevations are. You won't use this feature for the current task because you are not placing an inlet at a low point.

14. Right-click and then click End Run.

    The command will end, and the new inlet will be visible on the road.

15. Zoom in to the new inlet and use the editing gizmos to align it with the curb and set it at an elevation just above the pavement surface.

16. View the new inlet and pipe from underground and observe how it is connected to the inlet from the system you created previously. If you like, you can restore the bookmark named Drainage Network Underground.

    The software automatically applied depths and slopes based on design rules to produce a design that is practical.

17. The Drainage Design toolbar should already be expanded, but if it is not, click the Drainage Design icon.

18. On the Drainage Design toolbar, click the Analyze icon.

19. On the Analyze toolbar, click the Quantities icon.

20. Click the road you have been working on.

    The Quantities asset card will open showing you a tabulation of inlets, manholes, and pipelines for this road (see Figure 2.42). There are no culverts, so that value reads 0.

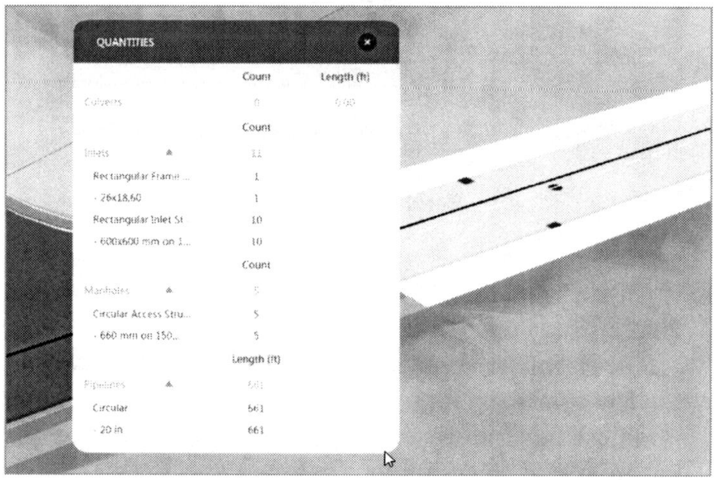

FIGURE 2.42 The Quantities asset card showing quantities of drainage components for a road

21. Restore the bookmark named Main Industrial Park Road.

22. With the Quantities asset card still open, click the long road in the view that runs north to south.

    The Quantities asset card updates to reflect the drainage items associated with this road. This includes a culvert and much higher quantities of inlets, manholes, and pipelines when compared to the previous results for the other road.

    You can view the results of successfully completing this exercise by selecting the Ex_2_9_End proposal.

Now You Know

Now that you have completed this chapter, you know how to create a watershed for a single point or to create one or more watersheds along a road. You can also analyze watersheds using the Watershed asset card to calculate flows. After calculating flows, you are able to create, modify, and analyze culverts. You know how to get InfraWorks to automatically create the culverts along a road as well as how to create them manually wherever you like. You also know how to design and analyze culverts using the Culvert asset card and how to export the analysis results for further use. Finally, you now have an understanding of drainage networks and can create, modify, analyze, and quantify them (and culverts).

Now that you have completed this chapter, you are able to begin designing drainage for your own projects using Drainage Design for InfraWorks 360.

Index

Note to the Reader: Throughout this index **boldfaced** page numbers indicate primary discussions of a topic. *Italicized* page numbers indicate illustrations.

D

E

F

G

H

I

L

M

N

O

P

Q

R

S

T

U

V

W

CPSIA information can be obtained at www.ICGtesting.com
Printed in the USA
BVOW01s1948290115

385624BV00003B/17/P

9 781118 915967